Insect Pests of Temperate Fruits Crops and their Management

The Authors

Dr. Asma Sherwani is presently working as Assistant Professor in the Division of Entomology, Faculty of Horticulture Sher-e-Kashmir University of Agricultural Sciences and Technology of Kashmir (SKUAST K) Shalimar, Srinagar. She did her post-graduation from P.D.K.V., Akola, Maharashtra. She qualified National Eligibility Test (NET) in Agricultural Entomology. After completion of Ph.D from SKUAST K, she started her career as an Assistant Professor. Dr. Asma Sherwani has 7 years of experience in research, teaching and extension in agricultural and horticultural entomology with specialization in Acarology. She has published more than 20 research publications in journals of repute on different aspects of entomology.

Dr. Malik Mukhtar Ahmad is presently working as an Assistant Professor senior scale in the Division of Entomology, Faculty of Horticulture Sher-e-Kashmir University of Agricultural Sciences and Technology of Kashmir (SKUAST K) Shalimar Srinagar. He qualified National Eligibility Test (NET) in Agricultural Entomology. After completion of Ph.D from SKUAST K, with specialization in pesticide residues, he started his career as an Assistant Professor at Faculty of Agriculture, Wadura Sopore. He is presently deputed to Research Centre for Residue and Quality Analysis (RCRQA). Dr. Malik Mukhtar Ahmad has 10 years of experience in research, teaching and extension in agricultural and horticultural entomology. He has published more than 25 research publications in journals of repute on different aspects of entomology and pesticide residues.

Insect Pests of Temperate Fruits Crops and their Management

– *Authors* –

Asma Sherwani

Malik Mukhtar Ahmad

2018

Daya Publishing House®

A Division of

Astral International Pvt. Ltd.

New Delhi – 110 002

Cataloging in Publication Data--DK
Courtesy: D.K. Agencies (P) Ltd. <docinfo@dkagencies.com>

Sherwani, Asma, author.
Insect pests of temperate fruits crops and their management / authors, Asma Sherwani, Malik Mukhtar Ahmad.
pages cm
Includes bibliographical references and index.
ISBN 9789387057678 (International Edition)

1. Fruit--Diseases and pests--Control--India--Kashmir, Vale of. 2. Insect pests--Control--India--Kashmir, Vale of. I. Ahmad, Malik Mukhtar, author. II. Title.

LCC SB608.F8S54 2018 | DDC 634.049709546 23

Published by : **Daya Publishing House®**
A Division of
Astral International Pvt. Ltd.
– ISO 9001:2015 Certified Company –
4736/23, Ansari Road, Darya Ganj
New Delhi-110 002
Ph. 011-43549197, 23278134
E-mail: info@astralint.com
Website: www.astralint.com

Prof. M.Y. Zargar
Director Research
SKUAST-Kashmir

Foreword

It is a matter of immense pleasure that Dr. Asma Sherwani and Dr. Malik Mukhtar have undertaken this initiative of bringing out this *Textbook on Insect Pests of Temperate Fruit Crops and their Management*. Temperate horticultural ecosystem is witnessing a major change with respect to the climate change and susceptibility of introduced fruit plant to the native insect pests, their diversity, feeding habits and host range. The spectrum of introduced and native insect pests of fruit crops in horticultural ecosystem keeps on changing as occasional pests are becoming regular pests thereby putting great pressure on integrated pest management in general and chemical intervention in particular. In such a scenario the knowledge of insect pest bionomics, life cycle and management becomes even more important.

The contents of this book provide a comprehensive account of various insect pests of temperate horticulture and language is lucid in nature. The contents of the book cover all insect pests of temperate fruit crops like pome, stone, nut and berry. The chapters focus on insect pest description, host plant range, general appearance, nature of damage and management. This book will provide useful practical information to students, teachers and extension workers.

This book has compiled all the insect pests of temperate horticulture that are present all over the world. The book has filled a big void, as there was no book available that could address insect pest problems in horticultural crops of Kashmir valley. The joint initiative of Dr. Asma Sherwani and Dr. Malik Mukhtar deserves appreciation for their effort in collection of literature on insect pests in a single manuscript and further adding material to the current literature.

I wish their endeavour all success.

M.Y. Zargar

Preface

Temperate areas of the world are home to popular horticultural fruits and nuts. The temperate fruits require cold season annually to maintain productivity. These crops are grown in those regions that fall between the tropics and the polar areas. The main temperate fruit growing regions in the world are Europe, North America and Asia. The temperate fruits occupy an important place in balanced human diet, as they are good source of vitamins and minerals. In India the main temperate fruit growing area is Kashmir province. The climate, soil and environment of Kashmir are conducive to temperate fruit production. Apple and other temperate fruits of Kashmir are relished worldwide for its texture and taste. Kashmir being part of south Asia has the unique distinction for temperate fruit production as geographically it is placed in Himalayan region of Indian sub-continent that is favourable for growth of apples and other temperate fruits since times immemorial. The total area of Kashmir valley is 100 km wide, which accommodate an area of 15,520 sq. km in area. Kashmir is home to several small minor Himalayan valleys like Anantnag, Baramulla, Budgam, Srinagar, Pulwama and Kupwara comprising approximately 85 miles in length and 25 miles in breadth with an average altitude of 5300 ft. above mean sea level.

There are still some places in Kashmir that gives us pleasure of its rich horticultural background like Tangdar (Pear orchards), Chare-van (Apricot forest), Chunt-var (Apple yard) and Badam var (Almond yard). Among the major challenges to the temperate fruit industry, insect pests are one of them. Apart from favourable climate and facilitating geography the production of temperate fruits of Kashmir lags far behind the other temperate areas of the world. One of the reasons for its low production are insect pests, which devastate the produce causing a considerable loss to state exchequer. The information about insect pests of temperate fruits of Kashmir was sporadic and there were gaps in information about the pests of fruits present in Kashmir. There was a dire need to publish a concise book that could serve a background to life cycle, biology and management of insect pests that could also contain some principles and practices of insect pest control. That prompted us to write this book where all information could be presented in a single manuscript.

This book has been written primarily for agricultural entomology students and educated farmers of temperate areas where they can find all information about insect pests of temperate fruit crops and nuts under one cover. This book covers almost all the insect pests of temperate fruits that are grown in tepid latitudes. This book contains chapters on insect pests and management of Apple, Pear, Plum, Peach, Almond, Apricot, Walnut, Chestnut, Strawberry, Raspberry, Blackberry, Mulberry, Grapes, Pomegranate, Cherry, Fig, Loquat, Quince Kiwi and Persimmon grown in Kashmir as well as in all the temperate belts of world.

We express our gratitude to Prof. Nazeer Ahmad, Vice Chancellor, SKUAST-K for his help and constant encouragement received during the compilation of this manuscript.

We are also thankful to Director Research SKUAST-K and Head Division of Entomology for their help and advice received during the write up of this manuscript.

Suggestions for improvement will be most welcome.

Asma Sherwani

Malik Mukhtar Ahmad

Contents

Chapter 1

Insect Pests of Apple and their Management

Apple *Malus domestica* (Borkh) is considered to be the world's oldest fruit crop. It is premier table fruit of the world and excels other fruits in having prolonged keeping quality and wide variety of flavor and taste. Apple is a highly remunerative crop and is grown in all temperate regions of the world. In India apple is grown in Jammu and Kashmir, Himachal Pradesh, Uttarakhand and some parts of Arunachal Pradesh. Jammu and Kashmir accounts for sixty per cent of total apple produced in India and is the most important fruit among all the cultivated fruits in the state. The state annually produces 13.73 lakh metric tonnes of apple fruit from an area of 1.28 lakh hectares. It forms the backbone of the state's economy and is main cash crop for the small and marginal farmers. However, the production and quality of apple is poor as compared to that of the developed countries because of several factors including insect pests. Although a large number of insect pests attack apple crops but some of them are very serious and need attention for their control. Besides the insect pests, mites are also associated with apple production and cause significance economic losses to commercial fruit growers. Insect pests found in apple orchards can be classified into two groups depending upon plant parts, which they attack. Direct pests are those insects that feed on apple fruits while indirect pests are those that attack leaves, trunk and other parts of the tree. Examples of direct pests are apple maggot, codling moth and other internal fruit feeders. Pests like leaf miner, aphid and mites are indirect pests since they do not directly injure fruits. They are also called major and minor pests when classified in terms of seriousness of their infestation. The major insect pests attacking apple fruit in temperate areas of the world are as follows:

1. San Jose scale

San Jose scale, *Quadraspidiotus perniciosus* Comstock (Diaspididae: Hemiptera) is widely distributed in the entire apple growing countries of the world. In India, it was introduced from France in 1906 and now has been recorded on more than 32 host plants.It has spread to every continent except Antartica. This is the most damaging pest in northwest India, particularly Kashmir and Kullu valley of Himachal Pradesh. The pest prefers to feed on plants belonging to the family Rosaceae such as apple and pear but can survive on other fruits of hilly region. Thus, more than 150 host plants have been recorded.

Host Plants

Apple, Pear, Peach, Plum, Cherry, Apricot, Walnut, Almond, Mulberry, Chestnut *etc.*

General Appearance

The adult female reproduces ovoviviparously *i.e.*, the eggs develop in an ovisac inside the mother to be born as nymphs. Each female give birth to 200 to 400 nymphs. The first instar nymphs move for about 12-24 hours and fix themselves at suitable places on host tree and begin to feed by sucking the cell sap. Simultaneously, they secrete a waxy material (white cap stage) over themselves, which give them the name scale insects. The waxy covering turns black and is known as the black cap stage. Later the covers turn various shades from grey to black. They become full grown in 3 to 4 days and within next 10 to 14 days the female again start giving birth to new ones. The longevity of gravid mother is 50 to 53 days and male adult develops into a winged adult in 25 to 31 days and live only 24 to 32 hours during which it fertilizes the non-winged female inside the scale covering. The insect overwinters in the nymphal stage (black cap stage) inside the scale covering. The insects are active from April to December. There are 6 to 7 overlapping generations in a year. The scale is disseminated by various birds, bats *etc.* The first instar crawlers are the main dispersal phase and are carried for a few kilometers by the wind.

Nature of Damage

It is serious pest of apple, pear and peach but can survive in several other temperate fruit trees like, almond, apricot, cherry, chestnut, plum, mulberry *etc.* The adult females and nymphs are destructive. They suck sap from twigs and branches as the result nursery plants become weak and die. The leaves, twigs, fruits and sometimes even the entire bark may be seen covered with ashy-grey scales which can be easily scraped off exposing the orange coloured individuals beneath. The affected fruits present pink coloured areas around the scales and the market value of such fruits is reduced.

Management

- ☆ Orchard sanitation greatly reduces the damage. Heavily infested branches should be removed and burnt. Shade trees especially willow, poplar should not be planted in and around fruit orchards.

Figure 1: Females with Scale Cover Removed to Show Scale Body.

- ☆ Several parasites and predators attack San Jose scale. The parasitoids recorded from San Jose scale include *Encarsia perniciosi* (Tower) and *Aphytis proclia* (Walker). Although they destroy many scales, they do not provide enough control to prevent damage. Natural enemies may become numerous in orchards that are not sprayed with insecticides, but even under these conditions biological control has not been adequate. Currently, biological control is only a supplement to chemical control. The parasite, *Encarsia perniciosi* (Tower) may also be released to check the overwintering population and have proved to some extent effective in controlling this insect pest.
- ☆ *Chilocorus bijugus* (Mulsant) or *Coccinnella septempunctata* (Linnaeus) @ 30 to 50 adults/infested trees of San Jose scale can be released during the month of July to September to suppress the pest.
- ☆ Spray diesel oil emulsion with fish oil soap (potash based) in the ratio of 1:7 (stock solution and water) *i.e.*, 6.35 per cent or spray horticultural

Figure 2: Apple Twig Infested with Scale Insects.

mineral oil (HMO) @ 2 per cent (2 litres of HMO in 100 litres of water) in early spring season when trees are completely defoliated and when weather is clear and temperature is above 4°C.

- Spray Chlorpyrifos 20 EC or Dimethoate 30 EC with dosage of 100 ml/100 litres of water during summer to kill the crawlers and newly settled scale insects. Select healthy plant from nursery and treat them with Chlorpyrifos 20 EC @ 100 ml/100 litres of water before planting in the field.
- Degree-day model is helpful for timing spray for crawlers in June. The lower and upper developmental thresholds of San Jose scale are 51°F and 90°F. The model should be started at first male scale capture in a pheromone trap (*i.e.* bio fix). Because male scale flight is difficult to monitor accurately in commercial orchards, the regionally established bio fix for codling moth is often used to start the San Jose scale model, as the flight of both insects commonly begins on the same day. If bio fix is not available, start the model at full bloom of red delicious variety of apple. Apply pesticide sprays aimed at crawlers between 400 and 450-degree days after bio fix. Control of infestation in the early stages will not only protect tree vigor but will prevent them from spreading to other trees in the orchard.

2. Woolly Apple Aphid

Wooly aphid, *Eriosoma lanigerum* Haussmann (Aphididae: Hemiptera) is native to America and is cosmopolitan in distribution except in hotter parts of the tropics. The pest is active throughout the year. The nymphs secrete wooly filaments of wax over their bodies, hence the name wooly aphis.

Host Plants

Apple, Crab apple, Almond, Pear and Quince.

General Appearance

The aphid reproduces both sexually as well as parthenogenetically, of which the latter is more common. The pest reproduces throughout the year except in the colder months from mid-December to mid-February. The wingless forms are present all through the years whereas the winged forms are seen only from July to October. From March onward, each female produces 30 to 116 nymphs parthenogenetically. The nymph within 24 hours starts secreting woolly filaments and wax over their body hence named as woolly aphids. Four nymphal instars lasts for 11 days in summer and 93 days in winter. Both winged and wingless adults are formed. With the onset of winter sexual forms appear. Male and female mate to produces eggs. These eggs remain dormant till the arrival of spring. During winter the nymphs already present on the tree migrate downward to enter the root for hibernation. After the hibernation period is over *i.e.*, April onward the nymphs from roots moves upward on tree branches to complete the life cycle. The dormant eggs also hatch into nymph on the arrival of summer. About 13 generations are completed in a year. Maximum multiplication of the pest occurs in summer and early monsoon.

For dispersal winged adults fly away to new host plants, while wingless forms are blown off by the wind.

Figure 3: Wooly Aphid Colony.

Nature of Damage

Wooly aphid is one of the most destructive pests of apple throughout the world. Both adult and nymph suck the juice from the bark of the trunk and from the roots of the host plant, as a result infested twigs shrivel. It attacks primarily the underground roots, which develops swelling and the whole plant may die. The winged form of the pest attack trunk branches and fruit stalks *etc.* The pest remains active throughout the year except in cold months of December and January. Above the ground, infestation is characterized by the presence of cottony patches scattered over the stem and branches. Nursery plants are severely affected.

Management

- ☆ Use resistant root-stock like Golden Delicious, Northern Spy and Morton Stock 778, 779, 789 or 793 has been found effective to prevent damage by aphid.
- ☆ The aphid population can be effectively checked in the month of July by an exotic parasite *Aphelinus mali* (Haldeman) @ 1000 to 1500 or *Coccinnella septempunctata* (Linnaeus) @ 15 to 30 per infested tree.
- ☆ Spray Dimethoate 30 EC @ 100 ml in 100 litres of water during summer months.
- ☆ Avoid insecticidal sprays where parasite *Aphelinus mali* (Haldeman) is present.
- ☆ Spray Methyl-*O*-demeton 25 EC @ 80ml/100 litres of water during late spring will prevent infestation in summer. For controlling the root forms, apply Carbofuran 3 CG @ 70 to 100 g/tree under the canopy area followed by hoeing of soil.

3. Fruit Fly

Apple maggot *Rhagoletis pomonella* Walsh (Tephritidae: Diptera) was discovered in Oregon USA in 1979. Heavily infested fruit is inedible and is suitable only for cider or animal feed. It is very important to manage this pest to protect apple crop.

Host Plants

Apple, Cherry, Pear and other fruit trees.

General Appearance

The eggs are usually deposited singly just beneath the skin of the apple. Eggs hatch after 2 to 10 day incubation period, depending on the ambient temperature. The maggots pass through three instars, spending 20 to 30 days feeding within the fruit. The maggots develop more rapidly and mortality is lower in early maturing, soft cultivars than in firmer fleshed, later-ripening apples. Upon completing their third-instar feeding, the maggots drop to the ground, burrow into the soil, and molt to a fourth instar, which is quickly followed by another molt to the pupal stage. Pupae are found within puparia made from the third-instar skin. The majority of pupae are located within 50 mm of the soil surface. Pupae pass the winter in diapause. The adults emerge from the soil from mid-June to early July. Peak emergence occurs during mid- to late July and is usually completed by the end of August. Emergence patterns vary considerably among different geographic locations and even within a specific area, depending on the host and environmental parameters, particularly temperature, soil type, and rainfall. Female flies are black, have a pointed abdomen with four white cross bands. The males are smaller and have three cross bands on a rounded abdomen. Wings are clear and marked with characteristic black bands. Newly emerged flies are sexually immature and spend considerable time on apple leaves feeding on honeydew excreted by aphids and other insects. The flies mature sexually 7 to 10 days after emergence and congregate on the fruit, where mating occurs. After mating, the female punctures the apple skin with her ovipositor to lay eggs. Females can lay an average of about 300 eggs over a 30-day life span.

Nature of Damage

Female flies make cavities on fruits with their ovipositor and lay 3 - 8 eggs in each cavity. On hatching, maggots feed on the pulp. The affected fruits gradually rot and fall down. Other worms, especially the larvae of the codling moth that infest the insides of apples are often confused with the apple maggot. However caterpillars often feed on the apple core while apple maggots feed on the fruit flesh. In infested fruit, the larvae are often difficult to detect due to their pale, cream colour and small body size.

Management

- ☆ Sanitation can help reduce apple maggot populations. Frequently pick up and remove any apples that fall during the growing season and after harvest. Place these apples in the trash. Do not compost them in your farm.
- ☆ Bagging is one of the methods, which can prevent fruit fly damage. After thinning the fruit, in early to mid-June, enclose each apple in a plastic sandwich bag, either a zipper closure bag or a plain bag closed with staples. Using a pair of scissors, snip the bottom corners of each bag, leaving a

small opening for water to run out. At harvest, remove the bag. Although bagging fruit may take a few hours, but the apples are protected from apple maggots for the rest of the season.

- ✰ The trap-out method uses sticky traps to capture apple fruit fly females that attempt to lay eggs on the fruit. Apple fruit fly traps are red spheres coated with tangle foot, a sticky substance that adheres to almost any surface and permanently holds insects. The flies are attracted to the red colour of the spheres, land on them, and are stuck. Hang traps in the trees by the end of June, to catch the apple fruit flies as they first attempt to lay eggs. Remove any leaves or fruit touching the traps.
- ✰ Since apple maggot spends most of its life cycle protected within the fruit or buried in the soil, the insecticides must be timed to coincide with adult fly activity. Apple fruit flies are active from late June to October. Apply Dimethoate 30 EC @ 100 ml/100 litres of water within 7 days of trapping the first adult on yellow sticky cards. Repeat applications every 7 to 14 days until pre-harvest or more frequently if it rains. It is not necessary to reapply an insecticide if no more apple fruit flies are captured on traps after 3 to 4 weeks.

4. Codling Moth

The codling moth, *Cydia pomonella* Linnaeus (Tortricidae: Lepidoptera) is a serious pest of apple. The pest has been reported from Europe, USA, Canada, South Africa, Australia, New Zealand, Afghanistan and Pakistan. In India, this pest is restricted to cold arid region (Ladakh) of Kashmir state. This pest is supposed to have entered into Ladakh from the North West frontier province of Pakistan where it was reported as a serious pest on deciduous fruits. In Ladakh, it is widely distributed in all the fruit growing areas and has been recorded from Karkitchoo, Mangbore, Hardass, Sainikund, Batalik, Dah, Hanoo area of Kargil district and Lamayuru, Scrubachan, Khaltse, Tamisgam, Basgo, Saspole and Nimmu area of Leh district. The four local varieties of apple grown in Ladakh behave differently in susceptibility to codling moth attack. However, a number of varieties have been introduced in Ladakh, which showed varying degree of susceptibility to codling moth.

Host Plants

Apple, Pear, Walnut, Peach and Apricot.

General Appearance

The adult emergence takes place in the month of June. The female moth after mating starts egg laying on fruits leaves and twigs. Moths are greenish to dark brown with chocolate-brown or copper coloured circular markings near the tip of forewings. The colour pattern resembles bark of the tree trunk, which makes the moths quite inconspicuous. The eggs being flat and white transparent in colour. The hatching takes place after 7 to 15 days depending upon temperature and humidity. Full-grown caterpillars are 16 to 22 mm long and pinkish in colour. Egg, larval, and pupal periods are 4 to 12, 28 to 35 and 8 to 14 days respectively. The newly hatched larvae bore through the fruit surface generally from the calyx end and feed near the surface for a time before boring. The larvae then bore into deep and feed on pulp and developing seeds until fully grown. Early stage larva being creamy white in colour, which later becomes light pink. Matured larvae come out from exit hole of the fruit and spin a cocoon in loose bark, cracks or on debris and may or may not pupate. The codling moth overwinters as full-grown larva within cocoon under loose bark, leaf litter, or any other sheltered place nearby. Pupation takes place during second fortnight of May, which ranges from 10 to 20 days depending upon temperature and other climatic condition. When spring comes, the larvae become pupae inside these cocoons and the moths emerge from the cocoons during March to April. The pest complete one to two generations in a year, depending upon the temperature and other weather parameters.

Nature of Damage

Eggs are laid singly on leaves, blossoms and fruits. The freshly hatched caterpillars feed on leaves for a while, then burrow inside the fruits at the calyx end and feed on the pulp. The entry holes become quite conspicuous as these are filled with dry brown frass and are surrounded by a dark reddish ring. The infested apples become brighter in colour than those that are not infested and also ripe prematurely. The fruits that are attacked early in the season often drop down before the crop is ready for harvest.

Management

- ☆ Strict domestic quarantine is to be followed by screening of consignments of fruits to prevent the spread of the insect from Ladakh to other apple growing regions. Collect and destroy the infested fruits to prevent the carryover of the pest. All the debris, weeds and loose bark should be removed from old trees and orchard to prevent the hibernating larvae to find shelter.
- ☆ Band the trees with grass ropes or sac (Jute) cloth in 3 to 4 folds before the larval descend to the ground for hibernation and the larvae should be collected and destroyed.
- ☆ Weekly release of egg parasitoid *Trichogramma embryophagum* (Hart.) at the rate of 20,000 adults/50 trees/week should be undertaken from first fortnight of June to end of August.
- ☆ Mating disruption is a relatively new codling moth control tactic that can be very effective in certain orchards. In this technique, the female sex pheromone of the codling moth is released in enormous quantities in the orchard, preventing the male moth from locating the female for mating. This technique is safe for non-target insects and don't leave any residue on the fruit.
- ☆ The trees should be sprayed with Chlorpyrifos 20 EC @ 100 ml/100 litres of water, or Dimethoate 30 EC @ 100ml/100 litres of water at the time of complete petal fall stage when the moths start emerging. First spray called

the calyx spray which should be applied at the time of the fall of petals and closure of calyx and followed by 3-4 cover sprays with Methyl-*O*-demeton 25 EC @ 80 ml/100 litres of water at an interval of 15 days. The spraying should be done two months before the fruits are plucked.

5. European Red Spider Mite

European red spider mite, *Panonychus ulmi* Koch (Tetranychidae: Trombidiformes) has been introduced from Europe around the year 1911. This spider mite has become one of the most important pests of all fruit growing belts of the world and considered by many growers to be most important pest, which is sometimes difficult to control.

Host Plants

Apple, Walnut, Almond, Vines, Mulberry, Blackberry, Peach, Pear, Plum, Prune, Cherry and Chestnut.

General Appearance

Adult female European red mites are bright to brownish red with an elliptical body. Adult females are larger than adult males. Females have four rows of curved spines on their backs, each spine borne on a whitish tubercle. These spines can be seen with a hand lens. The overwintering eggs are bright red, spherically shaped and have a distinct white stalk about as long as the diameter of the egg. This white stalk can be seen only under magnification. Summer eggs are translucent. European red mites overwinter as eggs on twigs and smaller branches of trees. Eggs begin to hatch just before bloom. The mites crawl to the leaves and suck the plant juices. After mating, females lay eggs. These eggs develop into male and female mites. Eggs from unmated females develop into males only. Eggs hatch into a six-legged larval stage. The larvae then pass through an eight-legged protonymph and deutonymph stage before becoming adults. Ordinarily, populations of the European red mite build up slowly during the spring, but under favorable summer conditions they can increase to unacceptably high populations. Hot, dry weather is favorable to increase population of this mite. Nine or more generations may develop per year.

Nature of Damage

European red mite injures the tree by feeding on leaves destroying chlorophyll, and increasing respiration. This is accomplished by insertion of the mite mouthparts into the leaf cells to withdraw the contents. All motile stages feed on the foliage, preferably on lower surface of leaves, but both leaf surfaces are attacked when populations are high. Characteristic brown foliage, starting as a subtle cast to the green leaf, which becomes bronze in severe cases, results from heavy mite feeding. Damage is more severe when mite infestation is followed by Alternaria fungus (*Alternaria mali* Roberts). High population of mites during late season can cause further indirect downgrading of fruit by depositing overwintering eggs. The extent of mite injury is influenced by numerous factors: (1) time of the growing season when injury occurs, (2) duration of feeding when injury occurs, (3) vigor and tree cultivar, (4) crop load, and (5) weather conditions. Extensive foliage injury may reduce the quality and quantity of fruit and the next year bloom. Excessive foliage injury early in a growing season is most detrimental to the trees. Injury at this time may result in excessive fruit drop, and it may reduce following years bloom. Midseason injury is less detrimental but with other stress factors may result in fruit drop, reduced fruit color, and reduced effectiveness of growth regulating sprays.

Management

- ✰ Up keep the vigour of the plant by applying balanced dose of fertilizers and proper irrigation. Remove branches carrying eggs of mites in autumn when leaves fall.
- ✰ Pruning is important to remove heavily infested twigs or small branches during winter. It also allows the better penetration and coverage of oils and pesticides. The pruned wood must be burnt immediately.
- ✰ Spray of HMO @ 2 per cent (2 litres in 100 litres of water) in March when buds are breaking will give some control of winter eggs and act as a first line of defense. Summer sprays of HMO @ 0.75 per cent (750 ml in 100 litres of water) are also effective.
- ✰ The predatory phytoseiid mites *Typhlodromus occidentalis* (Nesbitt) and *Amblyseius finlandicus* (Oudemans) are the most important predators in most areas, although in some of the more humid areas, *Typhlodromus pyri* (Scheuten) can also play a role. *Stethorus punctum* (LeConte) a small black Coccinellid (ladybird) beetle can consume large numbers of mites but is not attracted to the orchard until the infestation level is quite high.
- ✰ Malathion is of short persistence but may be useful if kill is required close to harvest. Spray late in the evening or at night for better coverage and contact. Closely monitor mite populations.

Figure 4: Webbing of *Panonychus ulmi*.

Figure 5: Adult of *Panonychus ulmi*.

- ☆ Other sprays include Dicofol (18.5 EC) @ 1.08ml/litre, Propargite (57 EC) @ 1ml/litre and Abamectin (1.8 EC) @ 0.55 ml/litre of water as summer spray. Fenazaquin (10 EC) @ 0.4ml/litre of water should be applied at post bloom stage of crop. Spraying of Dimethoate (30 EC) @ 1ml/litre of water to apical portion also control the mites.
- ☆ Encourage natural enemies into the orchard by using insecticides and miticides that are more selective towards insect pests. Use economic threshold levels to determine the necessity of spraying.

6. Green June Beetle

Green June beetle *Cotinis nitida* Linnaeus (Scarabaeidae: Coleoptera) commonly known as the green June beetle, June bug or June beetle. The green June beetle is active during daylight hours. The adult is usually 15 to 22 mm long with dull, metallic green wings; its sides are gold and the head, legs and underside are very bright shiny green.

Host Plants

Grapes, Peach, Raspberry, Blackberry, Apple, Pear, Quince, Plum, Prune, Apricot, and Nectarine.

General Appearance

Green June beetles overwinter in the soil. Adult emergence occurs during July and August. Green June beetles are often most abundant a few weeks before and during harvest. There is one generation per year. Mating occurs in the early morning. A strongly scented milky fluid secreted by the female attracts the male. Mating last only a few minutes after which the female enters her burrow or crawls under matted grass. Once the mating process has taken place, the female will lay between 60 and 75 eggs underground during a two-week period. The eggs, when first laid, appear white and elliptical in shape, gradually becoming more spherical as the larvae develop. The eggs hatch in approximately 18 days into small, white grubs. The grubs will grow to about 40 mm and are white with a brownish-black head and brown spiracles along the sides of the body. The grub will molt twice before winter. The fully-grown grub colour is glassy yellowish white shading toward green or blue at the head and tail. The grub has stiff ambulatory bristles on its abdomen, which assist movement. The grub normally travels on its back. The underground speed is considered more rapid than any other known genus of Scarabaeidae.The larvae feed largely on humus and mold but can do considerable damage to plant root systems. The insect is considered more injurious in its immature stages than as a beetle. Pupation occurs after the third grub stage, which lasts nearly nine months. The pupal stage occurs in an oval cocoon constructed of dirt particles fastened together by a viscid fluid excreted by the grub. The pupa is white when first formed but develops greenish tints just before emergence. The adults begin to appear in June after 18 days of the pupation period. The adult is from 15–22 mm in length and 12 mm in width. The colour is from dull brown with green stripes to a uniform metallic green. The margins of the elytra vary from light brown to orange yellow. The adult beetle will feed upon a variety of fruits.

Nature of Damage

Adult green June beetles may injure foliage of apple plants. Adults attack the fruit by chewing irregular holes in it, and in some cases the fruit is nearly devoured. Apple foliage is skeletonized so that all the leaf tissue between the veins is removed. The grub feeding on the roots occasionally kill orchard grass.

Management

- ✰ A type of digger wasp, *Scolia dubia* (Say), loves to parasitize June beetle grubs by laying eggs that hatch and feed on the grubs. Sometimes called the blue-winged wasp, these wasps are most active in early fall, usually flying around low to the ground in search of grubs. To encourage natural predators, like wasps, avoid the use of broad-spectrum chemical pesticides.
- ✰ If necessary, apply Carbofuran 3 CG @ 100 g/tree under the canopy area in the fall to control green June beetle grubs. During the adult flying season, pesticide soaked over ripe fruit can be placed around the perimeter of orchards to help control adult beetles.

7. Green Apple Aphid

The green apple aphid, *Aphis pomi* De Geer (Aphididae: Hemiptera) is widely distributed in the apple growing areas and first appears in apple orchards in late May to early June. The insect sucks sap from leaves and succulent terminal growth. Green apple aphids are usually found close to major veins on the underside of the leaf.

Host Plants

Apple, Pear, Peach, Plum, Quince and Crab apple.

General Appearance

Green apple aphids overwinter as eggs on the bark of the previous season's growth usually on rough areas, such as leaf and pruning scars, spurs, or terminals. Eggs begin to hatch around silver tip stage. These early spring nymphs, all females, are known as stem mothers. Stem mothers mature and feed on the opening buds and developing leaves then give birth to living nymphs without mating (parthenogenesis). These nymphs feed, mature, and produce living nymphs. Generation follows generation in this manner. Winged aphids soon appear, and for the remainder of the summer the population consists of winged and wingless parthenogenetic females that produce living nymphs. The winged aphids fly to other apples or other host plants and reproduce. In the fall, winged aphids produce nymphs that develop into true sexual forms. After mating with males, females lay the overwintering eggs, which hatch the following spring.

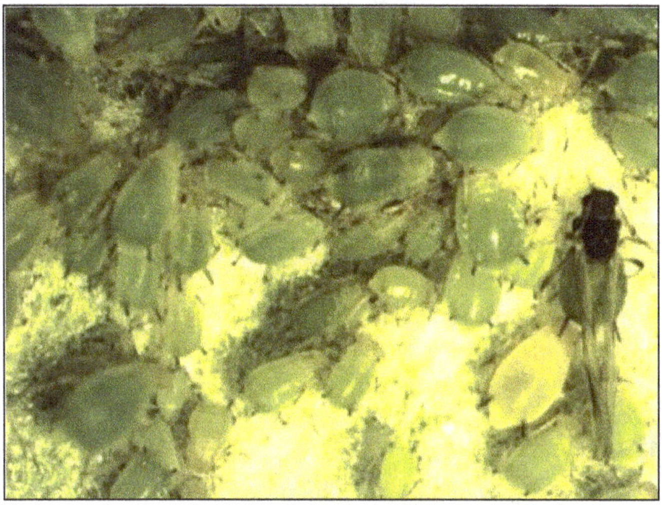

Figure 6: Apple Aphid.

Nature of Damage

Aphids such sap from leaves particularly near the growing tips. Heavy aphid feeding can cause curling and discoloration of developing foliage. Young trees may suffer reduced growth. The most extensive problems occur when high aphid populations develop on terminals and water sprouts and the area beneath is covered with honeydew. A sooty-mold fungus often develops on the honeydew, discoloring the fruit and adversely affecting its quality. Aphid feeding on immature apples results in stunted and deformed apples; their feeding on mature fruit produces russeted apples.

Management

- ✰ Aphid populations are affected by environmental conditions such as weather, plant health and natural enemies. If aphids do not have access to succulent new growth and they feed on older leaves, the number of nymphs produced drop by up to 50 per cent. If temperatures are 30-32°C and greater, females do not reproduce well. Aphids die when temperatures remain high for several days, and heavy rains wash aphid populations from leaves. A cool, wet spring favours aphid development and is unfavorable for the aphid's natural enemies.
- ✰ Green apple aphids are one of the few apple pests often managed by bio control agents. A number of beneficial insects are effective for biological control in apple orchards. The most commonly observed predator of green apple aphid include Lady Bird beetle *Coccinella septempunctata* Linnaeus, Green lace wing *Chrysoperla carnea* Stephens and syrphid flies. Insecticides applied against other pests can disrupt the natural enemy complex. When possible use pesticides less likely to disrupt predator populations.

- ☆ Manage nitrogen levels in plants to prevent excessive, lush terminal growth and help reduce aphid populations. Avoid summer pruning until terminal buds have set to prevent re-growth of shoots that are very attractive to aphids. Hand suckering in early June removes unnecessary vegetative growth that attracts green apple aphid.
- ☆ Insecticide treatments may be necessary in nurseries and young, non-bearing plants if populations of green apple aphid become high. Apply Horticultural mineral oil @ 2 per cent (2 litres/100 litres of water) in early spring.
- ☆ Spray Dimethoate 30 EC @ 100 ml/100 litres or Chlorpyriphos 20 EC @ 100 ml/100 litres of water during late May or early June. If infestation is heavy repeat the spray after 3 weeks.

8. Apple Tree Borer

Apple tree borer, *Apriona cinerea* Cheverlot (Cerambycidae: Coleoptera) is reported from India, Pakistan and Afghanistan. In India it is more commonly found in Kashmir, Himachal and Uttar Pradesh.

Host Plants

Apple, Peach, Fig, Pear *etc.*

General Appearance

Adult beetles begin to emerge from infested trees in late spring and continue to emerge until November. Female beetles prefer trees under stress or with damaged areas on the bark and deposit individual eggs in bark cracks or on the margins of wounds. Each female is capable of producing at least 100 eggs during her life. The grubs enter the bark after hatching and begin forming galleries. If the tree is vigorous, the grubs are often drowned by heavy sap flow. In weakened trees or in major bark wounds, the grub rapidly bores into the cambium layer and begins to form the gallery system. Development is completed in 1 year with nearly mature grubs overwintering.

Nature of Damage

The grubs cause damage to trees by creating extensive galleries in the phloem of the trunk and main branches. The galleries are partially filled with powdery frass. Infested areas are often re-infested year after year. Location of active galleries is often indicated by the presence of frothy sap bleeding from cracks in the bark. The damaged areas often become depressed with splits developing in the bark. The borers usually attack the sunny side of the tree. The galleries can eventually extend around the trunk and girdle the tree.

Management

- ☆ Collection and destruction of beetle and grub is very effective method to reduce the pest population. Prune and burn all attacked shoots and branches during winter.

- ☆ Since the pest feeds inside the woody portions no insecticide spray can be effective. The only way to control this pest is to locate the feeding holes, clear the passage and inject poisons.
- ☆ Insert in the live holes cotton wicks soaked in Dichlorvos 76 EC @ 3ml/litre of water or celphos @ 1 tablet/hole and seal with mud. Locate live holes and inject with carbon disulphide or chloroform or petrol and seal with mud to kill the adults.

9. Apple Root Borer

The apple root borer, *Dorysthenes hugelii* Redtenbacher (Cerambycidae: Coleoptera) is confined to foothills of Himalayan range and is a serious pest of apple in Jammu and Kashmir and Kumaon hills.

Host Plants

Apple, Apricot, Cherry, Peach, Pear, Walnut *etc.*

General Appearance

Female lays eggs singly or in small clusters in soil. Eggs are 1.3 mm in size. Newly laid eggs are white with a tinge of yellow and become dark brown before hatching. Grubs feed on the root. Grub longevity is 3.5 years. Grubs are cruciform, yellowish-white in colour. Development period ranges between 3-4 years. The full fed grubs reach 80 mm length and 12 mm in width. The pupae are about 48 mm long and usually found about 20–30 cm deep in the soil. Pupation occurs in earthen cells inside soil. Adults begin to emerge in May. Adults live about 40 days. The adult beetle is chestnut red in color.

Nature of Damage

The leaves become small and the branches wither. Tree becomes shaky and may die. Grubs bore into the sapwood and move upward or downward in the trunk, depending on the time of year. The galleries are continually enlarged as the grubs feed. Mature grub may bore an inch or more into the wood. Adults feed on the fruit, bark, and foliage. Grubs either bore or girdle around the roots. However, they usually do not cause any significant damage.

Management

- ☆ Keep trunks exposed, free of vegetation, weeds or trash shading the trunks. Most injury seems to occur where tall grass or other debris help hide the adults when they lay eggs.
- ☆ To check the incidence of this pest, avoid dry sandy soils for planting apple orchards. Inter culturing in the orchard helps in killing of grubs. Use well-rotted FYM and mix thoroughly with soil around the tree.
- ☆ Once infestation has occurred, it is important to treat the tree basins with Phorate granules (10G) @ 100 grams or Carbofuran 3G @ 50 grams per tree in the soil followed by hoeing.

10. Apple Leafhopper

Apple leafhopper *Typhlocyba pomaria* McAtee (Cicadellidae: Hemiptera) is the most common and serious of the leafhopper pests. Apple leafhopper is native to North America and appears throughout fruit growing regions of the world. These minute insects, are plant feeders that suck plant sap from grass, shrubs, or trees. Their hind legs are modified for jumping, and are covered with hairs that facilitate the spreading of a secretion over their bodies that acts as a water repellent and carrier of pheromones. It is an economically important pest of apple. Its pest status relates to its injury to the leaves, excrement on the fruit, and nuisance to workers.

Host Plants

Apple, Cherry, Peach, Grape, Blackberry *etc.*

General Appearance

Apple leafhoppers overwinter as eggs in the bark of 1 to 5 year-old apple trees. Eggs begin hatching in spring. Young leafhoppers are tiny, whitish green and wingless and are usually found on the underside of older leaves. They move quickly, going sideways or forward. Adults are greenish white and 1/3 inch long. They fly readily when disturbed. Adults develop in June and are active for several weeks. During this period, they lay eggs in the petiole and veins of leaves. The eggs begin to hatch during June. Leafhoppers feed in the orchard upto the fall. Overwintering eggs are laid in young wood during September and early October. The insect completes two generations per year.

Nature of Damage

The apple leafhopper is a leaf feeder and does not directly attack the fruit. As sap is sucked from the leaves, green tissue is destroyed, causing leaves to become speckled or mottled with white spots. Besides injuring leaves, leafhoppers deposit numerous small spots of excrement on fruit, potentially reducing its quality.

Management

- ✰ Young leafhoppers are much easier to control than adults. Effective control of the first generation may directly reduce high populations of the second. The first generation is a better target since the hatch is fairly synchronous, and leafhoppers of the age vulnerable to insecticides are present at one time. Also, insecticides may be used at lower rates since less foliage is present during the first generation. Thorough coverage of upper and lower leaf surfaces is necessary and considered essential for effective control.
- ✰ Monitor nymphs of leafhopper from bloom to petal fall. If in an average 3 or more nymphs are detected per leaf use the targeted insecticide. Chemicals like Dimethoate 30 EC @ 1ml/litre, or Methyl-*O*-demeton 25 EC @ 0.8ml/litre of water should be used in the first cover spray.

11. Apple Leaf Miner

Apple leaf miner *Phyllonorycter blancardella* Fabricus (Gracillariidae: Lepidoptera) is found in greenhouses, home gardens and landscaped areas across the globe. Leaf miners are the larval stage of an insect family that feeds between the upper and lower surface of the leaves. Although damage can restrict plant growth, resulting in reduced yields and loss of vigor, healthy plants can tolerate considerable injury.

Host Plants

Apple, Pear, Quince *etc.*

General Appearance

The leaf miner overwinters as a pupa inside apple leaves on the orchard floor. Adult emergence begins around the 1/2-inch green tip stage and continues through bloom. Females lay approximately 25 eggs singly on leaves. They hatch in 6 to 10 days. Larvae go through two stages *i.e.*, sap feeding and tissue feeding. Sap feeding larvae are very small; they pierce plant cells and feed on sap. Tissue feeding larvae are larger, have well-developed mouthparts, and feed directly on tissue inside leaves. A complete generation requires 35 to 55 days. Leaf miners feed in orchards from early spring until leaves drop in the fall.

Nature of Damage

Apple leaf miners injure apple leaves by internal feeding or mining. Each mature mine reduces the leaf's green tissue by about 5 per cent. Mines buckle the leaf like a small tent; hence the name spotted tentiform leaf miner. Excessive mining combined with drought, mite injury, or foliar diseases may be quite damaging. Mines remain visible after the leaf miner has emerged or been killed by sprays. To determine if mines are active, it is necessary to open them and observe for larvae.

Management

- ✰ Remove the weeds that are host plants of leaf miners. Remove the affected leaves by hand, collect fallen leaves and burn them. Maintain plant health with organic fertilizers and proper watering to allow plants to outgrow and tolerate pest damage.
- ✰ The parasitic wasp *Diglyphus isaea* (Walker) is a beneficial insect that kills leaf miner larvae in the mine.
- ✰ The leaf miner can be controlled by different chemicals like Phosalone 35 EC @ 1.4 ml/litre, Methyl-*O*-demeton 25 EC @ 0.8ml/litre or Dimethoate 30 EC @ 1ml/litre of water.

12. Tent Caterpillar

The tent caterpillar *Malacosoma indicum* Walker (Lasiocampidae: Lepidoptera) is one of the most conspicuous and familiar insect pests in India. Their silky, white tents can easily be seen covering the tips of tree branches. The emergence of tent caterpillars in the spring is a serious nuisance. During heavy infestation they will migrate and feed on many other plants.

Host Plants

Apple, Pear, Apricot and Walnut.

General Appearance

The pest is active from mid March to May and passes the remaining 9 months of the year in the diapausing egg stage. The eggs hatch by about the middle of March when buds appear on the plants. The larvae live gregariously and soon after emerging each spins a silken nest at a convenient and sheltered place on the tree. As the caterpillars grow, the nest is also enlarged until it is 0.3 to 0.5 meter across. During the day, the caterpillar rest in their nests and at night they feed on the leaves. The larval stage lasts 39 to 68 days and when full-fed, they spin oval, white and compact cocoons, each about 25mm in length. The pupal stage lasting 8 to 22 days is passed inside these cocoons in some protected place. The moths begin to emerge sometime in the third week of May and continue to do so till the beginning of June. The pre-oviposition period lasts 1 to 3 days and the females lay eggs in broad bands around the branches. Each band may consist of 200-400 eggs. The moths are short lived and a female in captivity may survive for 3 to 5 days. The life cycle is completed in a year.

Nature of Damage

In case of severe infestation, the entire plant may be defoliated and subsequently the caterpillars may feed even on the soft bark of twigs. When there is serious infestation, 40 to 50 per cent of the apple plants in an orchard may be defoliated producing a poor harvest.

Management

- ✰ Tents may be easily removed by pruning. It should be possible to prune out most of the tents if they are not too numerous or too high in the air. Do this in the early morning or evening when the weather is cool and the caterpillars are still in their tents. If the tent is within reach, break it with a stick and apply insecticide directly to it.
- ✰ The egg masses can be rubbed off from branches while pruning during the winter. This is a preventative strategy, which will reduce the numbers of hungry caterpillars that hatch in the spring. The egg masses appear as a gray or brown frothy material, which gets hardened to look somewhat like Styrofoam.
- ✰ If pruning out tents does not reduce the population below an acceptable level, then the affected plants may be sprayed with the biological insecticide *Bacillus thuringiensis* Balsamo. It contains a naturally occurring bacterial toxin that acts as a stomach poison after it is eaten by the caterpillar.
- ✰ Chemicals used to control tent caterpillars, are highly toxic to the tent caterpillar's natural enemies. While their use drastically reduces caterpillar

numbers quickly, the longer-term effects are less predictable. Contact insecticides which are currently recommended for the control of tent caterpillar include Chlorpyrifos 20 EC or Quinalphos 25 EC @ 1ml/litre of water. Spraying is effective at evening hours, as the caterpillar return to the nesting area at night.

13. Apple Rust Mite

Apple rust mite, *Aculus schlechtendali* Nalepa (Eriophyidae: Prostigmata) is rarely an important pest of apple in India. Natural predators or pesticide sprays applied against other pests usually control it. However, rust mite populations can build to injurious levels in some situations.

Host Plants

Apple, pear.

General Appearance

The species overwinters as inseminated females in crevices on twigs and under bud scales. Often large clusters can be found under a single scale. These females do not reproduce in the year in which they are produced. Some chilling is required for them before they lay eggs. They emerge to feed principally on the undersurfaces of leaves as the buds begin to open in the spring. They lay eggs that hatch into immature mites, which rapidly grow through two instars. Male rust mites are produced by unfertilized eggs and females are produced by fertilized eggs. There are several generations per growing season. Development is more rapid in warm temperatures. Overwintering forms can be produced as early as mid July if foliage condition is poor, but by fall only overwintering forms remain. They seek sheltered sites to spend the winter. In cooler climates, apple rust mite populations reach peak once in midsummer.

Nature of Damage

Apple rust feed on leaves producing a silvery cast to the leaf in the early stages, which tends to get browner as the season progresses. Bronzing caused by apple rust mite is more finely textured than spider mite bronzing and lacks the stippling produced by spider mites. Rust mite damage sometimes causes leaves to roll lengthwise. Like other pests that affect the foliage, damage disrupts photosynthesis and the tree water balance. Excessive amounts of damage, with peak populations can reduce fruit growth. As well as affecting fruit size, rust mite feeding can cause premature terminal bud set. Rust mites can also feed directly on the skin of fruit, causing a tan russeting by damaging the epidermis. In some cases there may be upward rolling of leaf edges and extensive feeding may produce silvery white blotches on the upper surface. Severe infestations may lead to leaf fall. Damage can be particularly severe on young trees in nurseries and newly planted orchards.

Management

- ✰ The predatory mites *Typhlodromus occidentalis* Scheuten attack apple rust mites. Predators can increase their numbers early in the season by

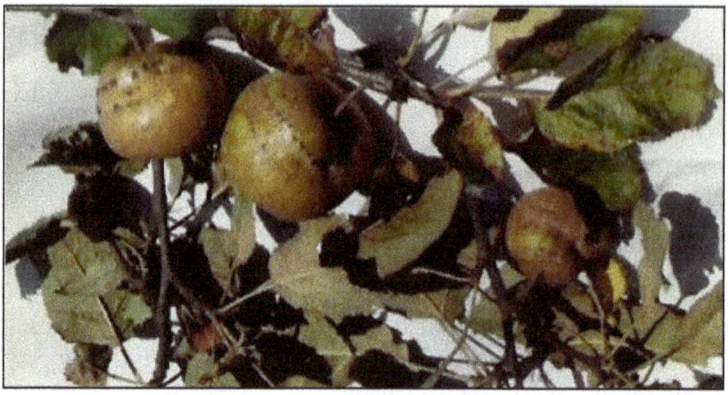

Figure 7: Rust Mite Damage on Apple Leaves and Fruit.

feeding on apple rust mites, thus providing better control of spider mite populations that develop later.

- ☆ Control measures for rust mites as foliage feeders are rarely needed. Populations of up to 50 mites per leaf in May or 250 in late June do not warrant control. Even if populations exceed 300 per leaf, the benefit of having them as an alternate food source for predatory mites must be weighed against the possibility of yield loss. An added benefit of rust mites is that their feeding preconditions foliage so that it is less suitable for build-up of the more damaging spider mites.
- ☆ Consider control only where rust mite populations remain high into July and there are no predatory mites. Make sure that trees are adequately irrigated during this period so that they are not subjected to water stress, which would exacerbate the effects caused by the mites feeding. An exception to the above strategy is in the case of high levels of rust mites pre bloom on Golden Delicious. If populations are high, then control to prevent fruit damage is warranted. Chemical controls will be most effective at about the pink stage of blossom development.
- ☆ Spray Fenazaquin (10 EC) @ 0.4ml/litre or Propargite (57 EC) @ 1ml/litre or Dicofol (18.5 EC) @ 1.08ml/litre of water in late June if the population of mite exceeds.

14. Two Spotted Mite

The two spotted spider mite, *Tetranychus urticae* Koch (Tetranychidae: Trombidiformes) feeds on a wide range of both wild and cultivated plants. Two-spotted spider mite is considered to be one of the most economically important spider mites. This mite has been reported infesting over 200 species of plants.

Host Plants

Apple, Cherry, Peach, Almond, plum, Pear, Raspberry, Mulberry, Chestnut and Grapevine.

General Appearance

Overwintering females are orange to orange-red. The body contents are often visible through the transparent body wall. They lay small spherical transparent eggs and many species spin silk webbing to help protect the colony from predators. The eggs are attached to fine silk webbing and hatch in approximately three days. The life cycle is composed of the egg, the larva, two nymphal stages (Protonymph and deutonymph) and the adult. The length of time from egg to adult varies greatly depending on temperature. Under optimum conditions spider mites complete their development in five to twenty days. There are many overlapping generations per year. The adult female lives two to four weeks and is capable of laying several hundred eggs during her life. The two spotted spider mite prefers hot, dry weather of the summer and fall months, but may occur anytime during the year. Overwintering females hibernate in ground litter or under the bark of trees or shrubs.

Nature of Damage

The mites generally live on the undersurface of leaves of plants where they may spin protective silk webs and they can cause damage by puncturing the plant cells to feed. The mite turns the leaves pale and causes the appearance of spots, later those turn yellowish bronze. Fruits also become underdeveloped. The mites feeding cause graying or yellowing of the leaves. Necrotic spots occur in the advanced stages of leaf damage. Mite damage to the open flower causes a browning and withering of the petals that resembles spray burn. When two spotted spider mites remove the sap, the mesophyll tissue collapse and a small chlorotic spot forms at each feeding site. It is estimated that 18 to 22 cells are destroyed per minute. Continued feeding causes a stippled-bleached effect and later, the leaves turn yellow, gray or bronze. Complete defoliation may occur if the mites are not controlled.

Management

- ☆ Up keep the vigour of the plant by applying balanced dose of fertilizers and proper irrigation. Remove branches carrying eggs of mites in autumn when leaves fall.
- ☆ Pruning is important to remove heavily infested twigs or small branches during winter. It also allows the better penetration and coverage of oils and pesticides. The pruned wood must be burnt immediately.
- ☆ The predatory phytoseiid mites *Typhlodromus occidentalis* (Nesbitt) and *Amblyseius finlandicus* (Oudemans) are the most important predators, although in some of the more humid areas, *Typhlodromus pyri* Scheuten can also play a role. *Stethorus punctum* LeConte, a small black Coccinellid beetle, can consume large numbers of mites but is not attracted to the orchard until the infestation level is quite high.
- ☆ The value of the delayed-dormant oil application aimed at overwintering females of mite has been proven effective. There is evidence on pear that delayed-dormant or pre-pink sprays suppress the overwintering females of two spotted mites, but this suppression does not prevent build-up later in the season. Overwintering females are physiologically and

Figure 8: Eggs and Adult of Two Spotted Mite.

morphologically different than the summer form, and chemical control of this form is more difficult. Specific miticides should be applied after the females lose their orange coloration and before these forms are produced again in the fall.

- ✰ When mite populations reach a density of 5 to 10 mites per leaf (80 to 90 per cent infested leaves) decide whether to use biological control or a miticide to prevent mites from increasing to higher densities. If neither predator is present at sufficient level for biological control to occur, nor mite populations are between 5 to 10 mites per leaf, apply a selective miticide.
- ✰ Spray Abamectin (1.8 EC) @ 0.55 ml/litre or Dicofol (18.5 EC) @ 1.08ml/litre or Propargite (57 EC) @ 1ml/litre of water.

15. Blossom Thrips

Thrips, *Frankliniella schultzei* Trybom (Cerambycidae: Thysanoptera) are tiny, slender insects with fringed wings. They feed by puncturing the epidermal layer of host tissue and sucking out the cell contents, which results in stippling, discoloured flecking, or silvering of the leaf surface. Thrips feeding is usually accompanied by black varnish like flecks of frass.They discolour and scar leaf, flower and fruit surfaces. Many species of thrips feed on fungal spores and pollen and are often innocuous.

Host Plants

Apple

General Appearance

These are small insects having a narrow, fringed wing with reduced wing venation and long marginal setae. Nymphs are elongated, white or yellow in colour. Females of thrips insert their eggs in flower tissue. Females of most plant-feeding species lay their elongate, cylindrical to kidney-shaped eggs on or into leaves, buds, or other locations where larvae feed. The embryonic stage lasts for four days and the first and second larval instars and two inactive and non-feeding stages *i.e.*, pre pupa and pupa occurs in the life cycle which take an average of 2.5, 2.5, 1.2 and 2.1 days, respectively. The adult male and female longevity is approximately 13 days. When the weather is warm, the life cycle from egg to adult may be completed in about 2 weeks.

Nature of Damage

Mites are associated with the blossom thrips and their population increases with the onset of the spring and flowering. Both nymphs and adults lacerate all floral parts and delicate unfolding leaves of vegetative buds. Consequently brown spots develop on affected plant parts. The lesions on androecia and gynoecia deteriorate the fruit quality and reduce fruit set and yield. The set fruits may drop. Heavily infested flowers bear sticky and faded appearance with indication of early senescence.

Management

- ☆ Avoid planting susceptible plants next to these areas, and control nearby weeds that are alternate hosts of pest thrips. Grow plants that are well adapted to conditions at that site. Provide appropriate cultural care to keep plants vigorous and increase their tolerance to thrips damage. Keep plants well irrigated, and avoid excessive applications of nitrogen fertilizer, which may promote higher populations of thrips.
- ☆ Predatory thrips, green lacewings, minute pirate bugs, mites, and certain parasitic wasps help to control plant-feeding thrips. To conserve and encourage naturally occurring populations of these beneficials, avoid creating dust and periodically do the dusting of plants, avoid persistent pesticides.
- ☆ Prune and destroy injured and infested terminals. Prune during specific times of the year to help control certain thrips. Ethion 50 EC @ 100ml/100 litres of solution with HMO used against San Jose scale will reduce the infestation of thrips.
- ☆ Thrips can be difficult to control effectively with insecticides, partly because of their mobility, feeding behavior, and protected egg and pupal stages. Improper timing of application, failure to treat the proper plant parts, and inadequate spray coverage when using contact poisons are common mistakes that can prevent potentially effective insecticides from actually providing control.
- ☆ Spray Methyl-*O*-Demeton 25 EC @ 0.8ml/litre or Dimethoate 30 EC @ 1ml/litre of water at pre bloom stage.

16. Shot Hole Borer

Shot hole borer *Scolytus nitidus* Muller (Curculionidae: Coleoptera) are small beetles that were introduced from Europe and have spread all over the temperate growing belts of world. In India Shot hole borer is distributed in Himachal Pradesh, Kashmir and Uttarakhand. On an average 5-10 per cent of apple trees get damaged annually by the attack of shot hole borer, which can increase upto 44 per cent in the unmanaged orchards during dry and hot weather conditions.The infested branches or sometimes the entire tree may be killed by this pest. The shot hole borer is a bark beetle that lives between the bark and the surface of the wood, scoring the sapwood. It feeds on the tree's succulent phloem tissue. The beetles bore into the wood of trees, forming galleries in which both adults and larvae live. All shot hole borers are attracted to injured or stressed trees.

Host Plants

Apple, Apricot, Peach, Plum *etc.*

General Appearance

The pest is active from April to September. Round pinholes are seen on the main trunk branches and tender shoots of the tree containing small blackish brown insects. The beetle is dark brown to black and tiny, with females between 1.78 mm and 2.5 mm long, and males even smaller, usually about 1.3mm long. Females bore through the trees bark, creating galleries under the bark. They plant the fungus in these galleries, where it grows and spreads throughout a susceptible tree. The female then lays her eggs in these galleries and when the eggs hatch, the larvae eat the fungus. The grubs develop into adults in about a month. Many more of the grubs develop into females than males, and the females mate with the males while still in the gallery. The females then pick up some of the fungus in their mouths, and leave through the entry holes created by their mothers to start the process again.

Nature of Damage

The grub and adult beetle tunnel into the sapwood and heartwood of the plant making small pinholes which cause die back of the branches. The females vector a fungus (*Fusarium euwallaceae*) that grows in their galleries, and the adult beetles and their larvae depend on it for food. While this fungus definitely helps the beetles out, it is really bad for the trees and clogs their water and food conducting tissues.

Management

- ☆ Pruning and destruction of borer infested branches during autumn. Birds, beetles, hymenopterous parasites and parasitic bacteria and fungi prey upon shot hole borers. Healthy, vigorous trees that are well cared for are less subject to attack. Give sick trees adequate water and fertilizer. In areas prone to sunburn or winter damage, paint trunks with a solution of equal parts of latex paint and water to prevent damage to the trees.
- ☆ Sanitation is the key to successful shot hole borer control. In winter, remove and burn infested and diseased trees or branches of orchard trees and

other nearby hosts. This will prevent beetle populations building up to the point where they might attack healthy wood.

- ✰ Baited traps like those used for monitoring can be used to control shot hole borers. Chemical controls can be applied whenever adults are present, from late March to September. However, they are of limited value, and the best control method is to keep trees healthy.
- ✰ Mud plaster infested trees with Carbaryl 50 WP and soil in the ratio of 1:6. In case of severe infestation spraying of Dimethoate 30 EC @ 100 ml or Methyl-*O*-demeton 25 EC 80 ml per 100 litres of water should be done for controlling the insect pest.

17. Stem Borer

The stem borer *Aeolesthes sarta* Solsky (Cerambycidae: Coleoptera) is a serious pest worldwide. The pest has spread from the Indian subcontinent to central Asia. In India the pest is widely distributed in Kashmir and Himachal Pradesh.

Host Plants

Apple, Cherry, Apricot, Peach, Pear, Plum and Mulberry.

General Appearance

Stem borer requires two years to complete a generation. Adults usually leave their pupal cells, located in the xylem, in April or the beginning of May, when the temperature averages 20°C. They are generally active in the evening and at night. During the day they hide under the bark, in larval tunnels, and in other refuges. After evening they leave their hiding places. The males appear first, and move about until morning on the same tree on which they developed. This species flies very little thus keeping natural spread relatively slow. The conspicuous feature of the adults is that the antennae are longer than body length. Adult maturation feeding has not been observed. Usually several generations develop on the same tree until the tree is eventually killed. Females lay eggs in slit-like niches in the bark of the trunk and the larger branches. Egg laying begins shortly after females leave their pupal cells and continues for about two months. Usually 1 to 3 eggs are laid per niche and females may lay a total of 240 to 270 eggs. The eggs hatch 9 to 17 days after being laid. Each grub makes its own tunnel, even if several eggs are deposited at the same place. Feeding begins in the cambium region and frass is ejected through the entrance hole. After feeding in the cambium for a period of time, the grub enters the xylem. At the end of their first season, grub make a long gallery that first goes upward for about 10 cm and then changes direction moving downward and resulting in a vertical gallery 15 cm long. At the bottom of this gallery, the grub overwinters, protected by a double plug made from borings. The following spring, the grubs resume feeding and construct tunnels deep into the wood. At the end of July, they prepare pupation cells protected by double plugs made from borings. Pupation occurs in these cells and about two weeks later adults emerge. Adults stay in the pupation cells, overwinter and emerge the following spring.

Figure 9: Stem Borer Adult.

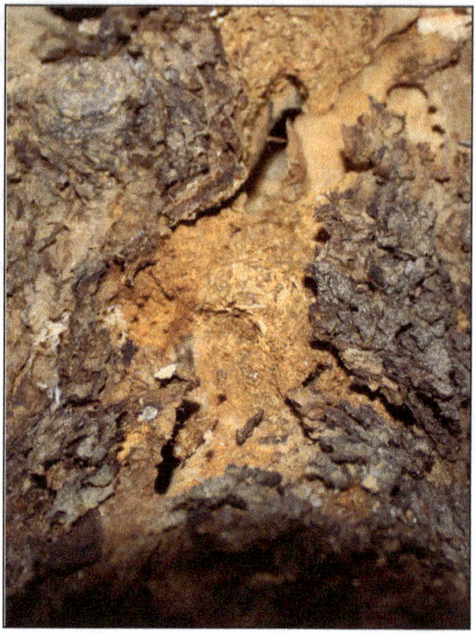

Figure 10: Apple Stem Borer Damage on Apple Branch.

Nature of Damage

The grubs damage the stem and branches by drilling big holes in the trunk. Saw dust is seen coming out from the holes. Young grub girdle a tree while feeding on the cambium, which leads to the rapid death of the tree. Young trees with a thin bark are the most susceptible to the beetle. Sometimes the presence of just 1 to 3 grub per tree is enough to cause mortality. Serious damage is also observed in shelterbelts and in fruit orchards.

Management

- Upkeep of the vigour of plant by applying balanced dose of fertilizers and proper irrigation. Collection and destruction of beetle should be done. Cutting and burning of infested plants should be done. Installation of light trap is helpful in collection of beetle. Prune the branches containing grubs before they entered the tree trunk.
- White washing of the uninfested trees should be done with slaked lime to prevent oviposition of eggs and protection from sunburn. Since the pest feeds inside the woody portions, no insecticide spray can be effective. The only way to control this pest is to locate the feeding holes, clear the passage and inject poisons.
- Insert in the live holes cotton wicks soaked in Dichlorvos 76 EC @ 3ml/litre of water or Celphos @ 1 tablet/hole or phenyl balls @ one ball/hole and seal with mud. Spray the tree with Dimethoate 30 EC or Chlorpyrifos

20 EC or Quinalphos 25 EC @ 1ml/litre of water during first week of May to prevent egg laying by newly emerged adults.

18. Gypsy Moth

Gypsy moth *Lymantria obfuscate* Walker (Lymantridae: Lepidoptera) commonly known as Indian gypsy moth or willow defoliator in Kashmir is distributed throughout the apple growing regions of the world. The first outbreak of pest occurred in 1889. These caterpillars multiply rapidly and can defoliate a large number of deciduous trees and shrubs in a short time. They are often seen on roadside trees and in neglected orchards. Besides defoliation the caterpillars produce large unsightly webs, or tents, in the crotches of tree branches. These webs are used to protect the caterpillars from predators.

Host Plants

Apple, Peach, Plum, Cherry, Apricot, Walnut, Willow and Poplar.

General Appearance

After being fertilized by a winged male, the female settles on the bark of a host tree for oviposition. It lay eggs during June to July, round, shining and light greyish brown eggs, in batches of 200 to 400. The eggs are laid under loose bark and are covered over with yellowish brown hairs. The eggs overwinter as such and hatch in March to April when the season warms up a little. The young caterpillar feed gregariously and as they grow through five instars. They become more mobile but remain together in large groups. The larval period is completed in 66 to 100 days. The pupal formation takes place in soil among debris and this stage lasts 9 to 21 days from May to mid July. The male moth lives for 4 to 10 days and female for 11 to 31 days. One generation is completed in a year.

Nature of Damage

Caterpillars feed on leaves. In heavy infestation trees may be completely defoliated. Repeated defoliation may weaken and kill most of the trees. The caterpillars of this pest defoliate the leaves and their feeding increases with the subsequent instars and the caterpillars feed voraciously on the entire tender leaves including the veins. The first instar caterpillars usually remain on the underside of leaves and are carried by wind from tree to tree, suspended by long threads that they spin. Caterpillars are nocturnal and gregarious in habit. They aggregate in large numbers on the ground under the fallen dry leaves near the base of the trees, crevices of bark or on lower parts of well-shaded main branches. After dusk the caterpillars start crawling through tree trunks and feed there. In severe attack the caterpillars defoliate the host plants completely thereby retarding the growth of the trees.

Management

- ☆ Keep your orchard as clean as possible. Remove discarded dead branches where the adult female moth is likely to lay eggs. Destroy all egg masses that are available for 8 months on tree trunks, branches and other hidden places. To check the infestation of this pest, destroy all the egg bands at the time of pruning in December to January.

1. The caterpillar hides in grasses grown at the ground level of tree. Therefore, orchard should be kept clean followed by collection and destruction of larvae. Burlapping of trees with gunny bags dipped in Chlorpyriphos 20 EC @ 100 ml/100 litres of water during May-June and hand destruction of larvae.

☆ The gypsy moth trap is used to monitor the moth population and may also prevent male moths from homing in on females. Spray Quinalphos 25 EC or Dimethoate 30 EC or Chlorpyrifos 20 EC @ 100 ml in 100 litres of water after petal fall will give satisfactory control of the pest.

19. Tortricid Moth

The Tortricid moth, *Archips pomivora* Meyrick (Tortricidae: Lepidoptera) is distributed throughout the world. In India it is found in Jammu and Kashmir and Himachal Pradesh. The larva is pest on cultivated apple and pear. They destroy young leaves flowers and occasionally newly set fruits. They also kill young lateral shoots.

Host Plants

Apple, Pear, Plum, Peach, Strawberry and Cranberry.

General Appearance

Moth passes through three generations annually with a partial fourth generation in some years. The moth has no winter resting stage. There is considerable overlap in the generations with major flight periods occurring during September to October, December to January, February to March, and April to May. Eggs are laid in clusters of 3 to 150 on leaves or fruit. A single female lays hundreds of eggs. Adults produced by the overwintering larval generation emerge during October and November. These give rise to the first summer generation, in which final instar larvae mature between January and mid February. Second generation larvae reach maturity during March and April, and the adults from this generation provide third generation eggs. Normally, the rate of larval development is slowed considerably during the winter, particularly when temperatures approach freezing thus the majority of larvae overwinter in the prolonged early juvenile phases of the second, third, and fourth instar. During this period they normally feed on herbaceous plants. The reinfestation of apple trees takes place during October to November, when moths of the third generation start laying eggs again on apple leaves.

Nature of Damage

The larvae cause significant damage to foliage and fruit. Early instars feed on tissue beneath the upper epidermis of leaves, while protected under self-constructed silken webs on the undersurface of leaves. Larger larvae migrate from these positions to construct feeding niches between adjacent leaves, between a leaf and a fruit, in the developing bud, or on a single leaf, where the leaf roll develops. The late stage larvae feed on all leaf tissue except main veins.

Management

- ✰ General insect predators and several species of spiders may influence the moth populations by feeding on eggs or larvae. High mortality has been reported during the initial dispersal of the newly hatched larvae. Appropriate sanitation practices during the dormant season can help prevent a buildup of tortricid moth. Mow broad-leaved weeds in and around the orchard before bud break. Remove mummified clusters when pruning.
- ✰ Spraying at 80 per cent of petal fall on Red Delicious variety will also give satisfactory control of tortricid moth. Dormant spray oils against San Jose scale will also give effective control of this pest. Control can also be achieved with the application of Dimethoate 30 EC @ 100 ml or Chlorpyrifos 20 EC @ 100ml/100 litres of water after petal fall.

20. Black Headed Fire Worm

The Holly tortrix moth, or black-headed fireworm *Rhopobota naevana* Hubner (Tortricidae: Lepidoptera) is distributed throughout the world. In India it if found in Jammu and Kashmir and Himachal Pradesh. The larvae destroy young leaves, flowers and occasionally, newly set fruits. They also kill young lateral shoots.

Host Plants

Apple, Pear, Cranberry.

General Appearance

Black headed fire worm overwinters in the egg stage on the tree. Eggs are laid singly on the back of leaves. They are flat, yellow disks and become very dark just prior to hatching. The black head of larvae can be seen inside the egg in the day or two before it hatches. The larvae have a distinct shiny black head and the body is of greenish to greyish colour and completes larval development in 2.5 to 5 weeks. Pupae are brown or yellowish in colour. The moth is small and flies in the daytime through dusk and mating occurs in late afternoon. Black headed fire worm completes two generations a year. In some rare years there may be a partial or complete third generation in warm years with early spring. Larvae hatch from overwintering eggs in spring and the resulting moths fly in June and July. The larvae of second-generation hatch during bloom and the resulting moths fly in late July and August.

Nature of Damage

First generation larvae feed on foliage, preferring the newly developing fresh growth. They prefer to feed in the growing tip, and web several leaves together to make refuges, as they get larger more terminals may be webbed together. They feed on the lower leaf surface leaving the upper surface intact, but this dies and turns reddish brown; the entire tip will turn brown as the larvae continue to feed. Second generation larvae feed on both foliage and fruit. When feeding on fruit, they feed at the fruit surface, causing a wound, which will be invaded by pathogens. They usually do not tunnel completely within the fruit.

Management

- ✰ Heavy prolonged rain near the time of egg hatch will kill many larvae. 24 to 48 hour flooding after the peak of egg hatch will kill a high percentage of larvae. If the weather is cool during this period and plants have not substantially broken dormancy, there will be no plant injury from such a flooding.
- ✰ Microbial insecticides based on *Bacillus thuringiensis* Balsamo have provided some control. This method would most likely be used if summer generation egg hatch occurred when bees were still pollinating the crop thereby restricting the use of conventional insecticides. Bt-based insecticides have short residual activity (24 to 48 hours) and therefore may require a second application. They should be timed to target younger larvae.
- ✰ Pheromone-mediated mating disruption has been developed for fruit worm and black-headed fire worm. Sprayable pheromone has been commercially developed. This method is acceptable for certified organic production.
- ✰ Some parasitic wasps attack black-headed fire worm, but per cent parasitism is always very low. Treat as soon as possible after egg hatch. Because of seasonal temperature changes, the spring egg hatch is easily overlooked.
- ✰ Quinalphos 25 EC or Dimethoate 30 EC or Chlorpyrifos 20 EC @ 100 ml in 100 litres of water after petal fall will give satisfactory control of the pest.

Chapter 2

Insect Pests of Pear and their Management

India produces all deciduous fruits including pome fruits (apple and pear) and stone fruits (peach, plum, apricot and cherry) in considerable quantity. These are mainly grown in the North Western Indian States of Jammu and Kashmir, Himachal Pradesh and in Uttaranchal. Pear *Pyrus communis* (Linnaeus) and other deciduous fruits were domesticated successfully in the early part of the 20th century, although some of them were reported to occur under semi-wild conditions much earlier. The pear is a perennial deciduous tree in the family Rosaceae, grown for its fruit. The tree is short deciduous with a tall and narrow crown and alternately arranged simple leaves. Pear trees can reach 9 m (30ft) and produces fruit for about 20 years. The quality of fruit and tree is affected by various insect pests, which are listed as under:

1. The Oriental Fruit Fly

Oriental fruit fly *Dacus zonatus* Saunders (Tephritidae: Diptera) is a very serious pest of stone fruits. It is known to attack more than 400 fruits and vegetables. The extreme adaptability of the pest gives an idea of the seriousness of the problem for its control. This fruit fly is native to Asia.

Host Plants

Apricots, Cherries, Citrus, Fig, Peach, Pear, Plum and Guava.

General Appearance

This fruit fly is active at temperatures above 15.5°C, and development of all life stages stops when temperatures dip below 5°C. The optimum temperature for activity (feeding, egg laying, *etc.*) and development is around 25 to 29.5°C. Adults have been spotted as early as late March and as late as mid-November. The fly has been reported to travel at least 15 miles in search of new hosts. Adults are relatively long-lived having three to nine overlapping generations per year. Adults need to feed on nectar, plant sap, and decaying fruit to mature sexually and for general survival. They normally feed during the morning and night is spent under foliage or any other protective crevice of hosts and non-hosts. The peach fruit fly mates typically at dusk, and lays eggs when immature fruit appears. The female puncture and make hole in the fruit to oviposit. A female can lay up to 150 eggs. The pre-oviposition period is about 20 days. Eggs are laid in a cluster of 2 to 9. Larva emerges out in 2 to 4 days. The maggots feed on fruit pulp. Larval life lasts for 4 to 16 days. Fully grown maggot fall on the ground to pupate in soil at a depth of 2 to 5 cm. After 7 to 9 days of pupation period adult emerges out. The adult pest remains active throughout the year, except during winter months however, their population remains maximum during rainy season.

Nature of Damage

After the eggs are deposited within the fruit, a sticky fluid normally seeps from this hole, eventually forming a brown, resinous deposit. The maggot feed throughout the fruit and cause premature fruit drop and significant yield reduction. Both adult and maggot are destructive. Adult makes puncture on fruit surface through which maggot enters inside the fruit and feed on the pulp. The infested fruit become undersized, malformed, show dark coloured patches, become rotten and subsequently fall.

Management

- ✰ Removal of fallen fruits. Each infested fruit can produce up to 400 fruit fly adults. Destruction of infested fruit is very import for fruit fly. Collected fruits should be buried 6 inches deep in soil. Young fruits should be completely bagged; bags must not have any holes in order to prevent oviposition.
- ✰ On eggs and larval stages Hymenopteran parasitoids are commonly employed; biological control alone does not provide high degree of control on sustainable basis. Traps are excellent tools for minimizing flies population.
- ✰ Spraying of Dimethoate 30 EC or Chlorpyrifos 20 EC @ 100 ml in 100 litres of water after emergence of adults provides effective control of insects.

2. Pear Psylla

Pear psylla *Psylla pyricola* Forster (Psyllidae: Homoptera) can be a limiting factor in pear production. It is a native species that produces abundant honeydew, which allows a sooty fungus to grow on the fruit surface. The result can be severe tree injury. It was first found in Connecticut in 1832 and within a few years it became a serious pest throughout all pear-growing areas in the world. As well as causing fruit russet, serious infestations can stunt, defoliate and even kill trees. The pest is a flush feeder means that the nymphs feed and develop primarily on the newer growth. During the growing season the leaves are hardened and development of pest can be limited to water sprouts only.

Host Plants

Pear.

General Appearance

The adults, which overwinter on trees or other sheltered places, become active anytime the temperature is above 40°F. Females begin laying eggs in late March and continue through the white bud stage. The eggs are creamy white when laid but turns yellow to orange as it develops towards hatching. One female can produce as many as 650 eggs. The peak of egg laying is green tip to green cluster bud. Egg hatch begins at the green cluster bud to white bud stage, with peak hatch occurring about petal fall. The nymph passes through five instars, each of which ends in a moult. Nymphs move to succulent stems and developing leaves to feed, with the heaviest concentration along the mid-veins of leaves and at the calyx end of fruit. The early nymphal stages produce more honeydew than the later stages. There are two adult forms, winter form and summer form. The first summer adults mature about 20 to 25 days past full bloom. They begin laying eggs on growing shoots as the population shifts from spur leaves to the more succulent shoot leaves. Late season infestations are typically found on water sprouts. Pear psylla passes the winter as adults in a state of reproductive diapause. They begin laying eggs when pear buds begin to swell. They deposit them around the base of buds and in other rough places on small twigs. After buds open, they lay eggs along mid veins and petioles of developing leaves, and on stems and sepals of blossoms. Adults continue to lay eggs through petal fall. This long egg laying period produces a wide age distribution of first generation nymphs, and some complete their development before the last have hatched. Summer form adults lay eggs on terminals, and the nymphs feed on leaves and stems of tender new growth. Some first generation nymphs feed on sepals and calyx ends of fruit. There are 2 or 3 generations of summer forms before the winter form generation develops in the fall. The pest is active from May to September.

Nature of Damage

Pear psylla damages pears in several ways:

- ☆ **Psylla shock:** In large numbers, pear psylla can stunt and defoliate trees and cause fruit drop. A carry-over effect may reduce fruit set the following year. These symptoms, called psylla shock, are caused by toxic saliva injected into the tree by feeding nymphs.
- ☆ **Fruit russet**: Nymphs and adults are phloem feeders. Honeydew, produced by nymphs, runs from fruit, causing dark russet blotches or streaks. This results in downgrading of fresh and sometimes processing fruit.
- ☆ **Pear decline:** Pear psylla also transmits a mycoplasma disease organism through its saliva. The disease damages sieve tubes in the phloem, which

prevents synthesized nutrients moving down the tree and results in root starvation. Trees suffering show decline and can recover if psylla density is low. The severity of the insect pest depends on the origin of the rootstock.

Management

- ☆ Minimize heavy pruning, which encourages the proliferation of terminal shoot growth. An overabundance of terminals provides more feeding sites for the psylla. Pear trees should receive the minimum amount of nitrogen fertilization necessary for proper tree and fruit growth. Over fertilization can cause extended terminal growth and delay hardening off, allowing optimal feeding conditions for the psylla. Remove water sprouts during late June and early July because water sprouts provide one of the only sources of succulent leaves at this time of the year, this technique can eliminate a large portion of the psylla population.
- ☆ To sample for pear psylla nymphs in the early season, examine at least 10 leaves per tree on a minimum of five trees per block. The action threshold at this time is 0.5 nymphs per leaf. For the summer generations again examine at least 10 leaves per tree on a minimum of five trees per block. The action threshold during the period is 1.5 nymphs per leaf. When the psylla population is primarily in the adult stage, examine the leaves for the presence of adult activity and egg laying.
- ☆ Several predators and at least two parasitoid species attack pear psylla. Common predators include *Crysoperla carnea* Stephens (green lacewing) adult and their larvae, *Coccinella septempunctata* Linnaeus (ladybird beetle) adult and grub. These predators feed on eggs and nymphs. Many species of spiders prey on adults. Most predators of psylla are general feeders and will prey on other pests in pear orchards, such as aphids and mealy bugs. The two parasitoids, *Trechnites psyllae* (Ruschka) and *Prionomitus mitratus* (Dalman) are small wasps and the adults lay eggs inside the bodies of the psylla nymphs, where the wasp larvae consume their hosts as they develop.
- ☆ In orchards with a history of psylla infestation, insecticidal control begins with a strong pre bloom spray program designed to eliminate as many overwintering adults as possible before they have the opportunity to lay many eggs. The first application should include horticultural mineral oil @ 2 per cent (2 litres in 100 litres of water), which has been shown to delay egg laying by overwintering adult. Typically, egg deposition and hatch occurs over a long period of time, making pesticide timing difficult. The early season oil application must be applied prior to egg maturation in the female psylla. With the oviposition period delayed, the delayed dormant spray at bud burst becomes extremely important because additional adult control can be achieved by waiting until adult psylla that are living away from the pear tree return from their overwintering sites.
- ☆ Light infestation may be controlled by spraying about 3 weeks after petal fall with methyl-*O*-demeton @ 80 ml/100 lit of water or dimethoate @

100 ml/100 lit of water or malathion @ 100 ml/100 lit of water. Heavy infestation may require two sprays, one at petal fall and another 3 weeks later.

3. Pear Rust Mite

Pear rust mite *Epitrimerus pyri* Nalepa (Eriophyidae: Prostigmata) is a common pest throughout pear growing areas and can cause serious fruit damage if untreated. While several predators feed on the pear rust mite, none controls it well enough to prevent commercial damage. Pear rust mites are so tiny that magnification lens is needed to see them, but the damage they cause is easy to see. These tiny creatures overwinter under leaf buds and loose bark. When temperature rises in spring, they emerge to feed on young, tender leaf tissue. When the tissue of the young leaves harden, the mites begin feeding on the fruit. Pear rust mite damage is only skin deep and comes off when you peel the fruit. The importance of this pest has increased where synthetic pyrethroid insecticides are used in pear psylla control that has reduced early-season populations of mite predators. Attempts to determine whether it can become established and reproduce on apple and related wild plants, such as hawthorn, have failed.

Host Plants

Apple, Pear, Cherry and Plum.

General Appearance

The mites cause leaf galls but in severe cases cause extensive damage to fruit buds and abnormal drop of blossom and immature fruit. Overwintering mites undergo a diapause during which they do not feed or lay eggs even though they may be active during warm winter weather. The mites lay eggs on basal bud scales. The pest overwinter as adult females beneath bud scales of leaf spurs and under loose bark of 1 to 2 year old twigs. As the weather begins to warm in early April, usually before buds break, mites move to developing clusters and begin feeding on the succulent parts of the buds. Eggs are produced shortly after mites become active. As buds open, adult and immature move to the expanding leaf tissue and eventually to fruit as the leaves mature and harden. Immature mites develop quickly through two instars, each followed by a resting stage. There are four to five generations in a year. During June and July, all stages are commonly found together on the lower leaf surface and around the calyx end of the fruit. By late summer, only females are produced and they soon move to protective over-wintering sites.

Nature of Damage

The most obvious damage caused by this mite is browning or rusting of the under surface of leaves. This is evident on leaves from the shoots of the current year. The damage of the pest becomes evident in the month of July. In some cases there may be upward rolling of leaf edges and extensive feeding may produce silvery white blotches on the upper surface. Severe infestations may lead to leaf fall. Damage can be particularly severe on young trees in nurseries and newly planted orchards.

Management

- The pear rust mite can damage fruit early in the season therefore sampling should begin at the cluster bud stage of tree development. The entire fruit cluster should be examined, including fruit buds or fruit, leaves and the younger green, woody tissue at the base of the bud. Later in the season, pear rust mite can be monitored using leaf samples taken to monitor pear psylla and spider mite. In addition, harvested fruit showing the injured fruit and extent of injury are helpful for detecting orchard sites that are at risk.
- Pear rust mite is attacked by several indigenous predators including predaceous mites Coccinellids, green and brown lacewings.
- Spray horticultural mineral oil @ 2 per cent (2 litres in 100 litres of water) at delayed dormant stage are effective for overwintering stages.
- While light infestations that do not cause significant damage can safely be ignored, heavily infested mature trees and young trees with severe leaf damage benefit from chemical rust mite control. Sulfur sprays can help bringing pear rust mites under control. Spraying during the pink or calyx stage best controls pear rust mites.
- Spray Clothianidin 50 WDG @ 14 g/100 litre of water or Cyenopyrafen 30 SC @ 30ml/100 litres of water. Fenazaquin 10 EC @ 40 ml/100 litres of water or Dicofol 18.5 EC @ 108 ml/100 litre of water at bud bursting stage are found effective against the pest.

4. Pear Leaf Blister Mite

Pear leaf blister mite *Phytoptus pyri* Pagenstecher (Eriophyidae: Prostigmata) is a microscopic gall mite, which affects the leaves of pears and some related trees. It is extremely small, 0.2 mm long, and is visible only with the help of a hand lens. However, the blisters are easy to spot. At first, they are about 2-5 mm in diameter, but they soon spread over to other leaves. They can cause problems on apple and pear orchards or backyard trees that have not been treated effectively with typical pesticides available to control other apple and pear pests like codling moth and pear psylla. Most common cases are in trees that have not been sprayed with any dormant oil treatments or which have been abandoned/neglected to the point where shoot growth in the upper and center portion of the tree is so dense as to make it nearly impossible to get penetration of any spray applied into the interior of the tree.

Host Plants

Pear and Apple.

General Appearance

The mite is easy to spot in field because of the characteristic damage it causes. It is often first seen on a few trees, or even branches, but can spread quite rapidly to other trees. The mite overwinter under bud scales then after emergence in spring their feeding cause cell atrophy and the development of blisters. The oviposition period is from March to October. Each female lays 7 to 21 eggs, which require 18 days to hatch. Sexual development requires 10 to 18 days. Total life cycle is completed in 23-36 days. There are 2 to 3 generations a year. Blister development takes place in 3 to 7 days. Small blisters have no opening and are without mites. The first damage is usually evident as rows of pink spots parallel to the midrib on folded leaves. Once holes appear in the blisters as the central cells die, the mites are able to invade them and subsequently population development occurs within these shelters.

Nature of Damage

The developing galls are green at first, becoming reddish, then dark brown or black. On the underside of leaves the blisters are corky and elevated. Often the blister mite probably does little harm to the tree, but in severe infestations the leaves may become severely disfigured, fall early and there may be damage to fruits. The mites overwinter under the outer scales where they feed on the developing buds.

Management

- ✰ If an outbreak is limited to a few leaves, pick them off before the blisters redden, and dispose off them. If the problem is widespread, removing the leaves is not practical. The mites do not have any serious adverse effect on tree growth or fruiting, but water stress should be avoided. Mulch trees in spring to conserve water in drought conditions.
- ✰ The mite is difficult to manage once it move into the blister it creates by feeding on the leaf tissue in the early spring. By petal fall, the mites will lay eggs and remain protected from predators within the leaf blisters.
- ✰ Delayed dormant oil applications also target the bud scales. Timing of spray during the season should be done prior to their movement into the leaf blisters after petal fall. The predatory mites, *Typhlodromus occidentalis* (Nesbitt), will feed on the exposed mites but do not provide economic control.
- ✰ Spray Dicofol 18.5 EC @ 108 ml or Fenazaquin 10 EC @ 40 ml in 100 litres of water provides effective control.

The other insect pests that affect pear are: San Jose scale, *Quadraspidiotus perniciosus* (Comstock). Codling Moth, *Cydia pomonella* Linnaeus Root borer, *Dorysthenes huegelii* (Redtenbacher), Tent caterpillar, *Malacosoma indicum* (Walker) Blossom thrips *Frankliniella Schultzei* Trybom and Shot hole borer and *Scolytus nitidus* (Muller) are discussed under the pests of Apple.

Chapter 3

Insect Pests of Plum and their Management

Plum, *Prunus domestica* Linnaeus is a deciduous tree in the family Rosaceae grown for its edible fruits and is cultivated in temperate areas worldwide for its fruit. The plum tree has an erect growing habit with a spreading canopy. It possesses large, thick, oval-shaped leaves, which are darker in colour on the upper surface than on the lower and which often have a serrated edge. The tree produces buds on terminal spurs on the branches with each bud generally producing 3–5 flowers. The fruit is a fleshy oval fruit with a single seed contained within a stone. The colour of the fruit varies with variety and fruits can be purple, blue, green, red or yellow. Plum trees can attain a height of between 6 and 10 m and can live for periods in excess of 50 years if properly maintained. Plum originates from Southwest Asia. Plum trees are afflicted with insect pests less often than other fruit trees. However there are few insect pests that can be truly devastating if given a chance. The insect pests, which affect plum, are as follows:

1. Plum Borer

The plum borer *Euzophera semifunerali* Walker (Pyralidae: Lepidoptera) have become a serious pest of plum since the introduction of mechanical harvesting which is associated with tree wounding. Pruning wounds, graft unions, burr knots, black knot and canker type diseases also serve as entry points for larvae on other tree and nut crops as well.

Host Plants

Plum, Almonds, Pecans, Walnuts, and Olives.

General Appearance

The female is slightly larger than the male and lay between 20-50 eggs during a relatively short 3-4 day oviposition period. Females lay their eggs singly or in small clumps nearby on the rough bark. The eggs are bright, reddish pink in colour when first laid but turn white with age and hatch in 7-10 days. The newly hatched larvae immediately enter the nearby wound to feed on the underlying cambium. The larvae pass through 7 instars. Reddish orange frass, webbing, and gum pockets indicate their presence. The plum borer overwinters as a nearly full grown sixth or seventh instar larva within a silken hibernaculum. The hibernacula are formed near the feeding sites on the underside of the overlying dead bark. The larvae become active in the spring as soon as the buds begin to open. Most larvae pupate within their overwintering hibernacula without additional feeding. First adult emergence generally occurs about two weeks later by the white bud stage and peak adult emergence occurs just after full bloom. The majority of the eggs are laid by petal fall, although adult emergence often continues for another three weeks. The first summer brood larvae emerge to feed on cambium and complete development by late May to early June. Adult emergence of the second brood generally begins by late May and peaks during harvest in mid-July. Adult emergence of this brood is often very long with many adults being caught in August and a few even to the end of September. This brood is the most devastating since emergence coincides with mechanical or manual harvesting. At this time, there is an abundance of fresh cracks and wounds in the bark that open the cambium for female oviposition. Second brood eggs hatch during harvest when chemical control is difficult because of pesticide residue problems on the harvested fruit. They live for 1 to 3 weeks. There are three to four generations each year.

Nature of Damage

Larvae bore into the tree leaving reddish orange frass and gum pockets. The boring is most damaging to the scaffold crotches or graft unions of young trees. Vigorous trees will heal over, but with heavy, prolonged infestations, scaffolds may break with wind or with a heavy crop.

Management

- ☆ Remove bark until live cambium is reached and look for larvae feeding along the edge of the live cambium or cocoons attached to the inside of the bark. Use forceps to look under fresh frass. Maintain tree vigour through proper planting, balanced fertilization, and adequate irrigation during drought periods. Avoid pruning before or during borer flight periods. Remove badly infested trees that serve as reservoir of infestation.
- ☆ Natural enemies may play an important role in reducing larval populations. Birds, especially woodpeckers, feed on larvae throughout the year. A number of species of parasitic wasps, predatory insects, and spiders also feed on the borer. Fungi of the *Hirsutella* spp. are the important pathogens, attacking larvae of plum borer and other lepidopterous borers.

- ☆ Apply insecticides only when borers are vulnerable. Monitor young orchards in spring and summer for frass and gum pockets. If larvae are present, spray trees from 1 foot above the scaffold crotch to 1 foot below, two to three times during the growing season.
- ☆ Spray Dimethoate 30 EC @ 100 ml/100 litres of water or Dichlorvos 76 EC @ 70ml/100 litres of water from mid to late April and subsequent applications at 6-week intervals to kill adults before they can lay eggs.
- ☆ Plug holes with cotton soaked in Dichlorvos 76 EC @ 300ml/100 litre of water and then seal holes with mud plaster.

2. Branch and Twig Borer

Branch and twig borer *Melalgus confertus* Le Conte (Bostrichidae: Coleoptera) is found worldwide wherever stone fruits are grown.

Host Plants

Peach, Plum, Nectarine and Apricot.

General Appearance

The beetle lay its eggs in the dead wood of a number of native and cultivated trees and shrubs outside the orchard or on dead plum limbs once an orchard becomes infested. Grub bore into the heartwood of the host and feed within this area for a year or possibly longer. Pupation occurs within the wood, and adults emerge in early summer. They often fly from native vegetation to orchards where they bore into small branches on the trees. There is one generation per year.

Nature of Damage

Grub bore into the wood at dead and dying parts of the tree. The adults bore into small twigs and branches, making round holes, commonly at the axil of a bud or fruit spur or at the fork of two branches. Feeding is often deep enough to completely conceal the adult in the hole. Close inspection reveals a hole or gnawed area in the crotch formed on shoot and spur. Branch and twig borers seldom cause economic injury.

Management

- ☆ The simplest way to control twig borers is to prune and destroy beetle-infested plant material. Flagging branches should be pruned at least a few inches before symptoms begin on a branch. If the remaining portion of the branch is hollow, prune again closer to the trunk until the center is no longer hollow. Pruned material should be burned. Practices to promoting tree vigour, such as watering, mulching, and fertilizing can aid the plant during recovery from beetle damage. Prevent sunburn and other injury that predisposes trees to damage.
- ☆ The many species of general predators found under the bark may assist in maintaining lower populations. Treatments with commercial formulations of the entomopathogenic nematode *Steinernema carpocapsae*

(Weiser), which can move through frass tubes to infect larvae, may be of some benefit.

☆ If good cultural controls are practiced chemical control is normally not necessary. Spray Dichlorvos 76 EC @ 70ml/100 litres of water or plugholes with cotton dipped in Dichlorvos 76 EC @ 3ml/litre of water and seal with mud plaster when the infestation is high.

3. Peach Tree Borer

Peach Tree borer *Synanthedon exitiosa* Say (Sesiidae: Lepidoptera) is among the most destructive pest of peach trees. The adult of this insect are blue black, clear winged moth, which unlike most other moths is active in the daytime. They tunnel and feed under the bark in living wood, destroying water and sap conducting tissues. This causes girdling, branch dieback, structural weakness, decline and eventual death of susceptible plants. Infestation sites also provide entry points for plant pathogens.

Host Plants

Peach, Cherry, Plum, Prune, Nectarine and Apricot.

General Appearance

The pest is active from April to October. The moths emerge from mid July to early September Adults are clear winged moths with blue-black bodies having yellow or orange bands across the abdomen. The female lays their eggs on the trunk, cracks and crevices and sometimes on the foliage of the plant. Eggs hatch in about 12 days and the young larvae emerge and bore into the bark trunk near the ground level. After overwintering in their feeding tunnels the larvae resume feeding in their old or new tunnels. Larvae of the peach tree borer are white with brown heads. The larvae mature during June or early July and later pupate in the tunnels or near the surface of soil before emerging as adults. Gum exuding from around the base of the trunk is evidence of peach tree borer. The pest completes its life cycle in one year.

Nature of Damage

This wood-boring insect can successfully attack healthy trees. The larval stage bores into the crown and trunk of the tree and mines the cambial layer making galleries under the bark. The main roots near the ground surface may also be attacked. If this occurs for several years, the tree may eventually become girdled and die. Older trees are more resistant to injury than young ones. The presence of the pest can be seen by the skin of pupae left attached to the bark of the trunk on the lower side or in the soil near the trunk base or sometimes on a larval tunnel under the tree bark.

Management

☆ Avoid injuries to trunks and roots. Protect tree trunks and branches from sunburn. Avoid pruning trees when borer adults are flying. Replace old declining trees. Monitor tree trunks and branches regularly to detect infestations before they become serious. Local infestations of bark beetles and other boring beetles on branches may be pruned out. If the main trunk is extensively bored, remove the tree and focus on protecting neighboring trees of the same species.

- ☆ Many tiny holes in tree trunks and branches may indicate borers. Large open tunnels with D or O shaped emergence holes filled with sawdust like frass indicate borer infestation.

  Apply Celphos @ 1 tablet/hole or phenyl balls @ 1 ball/hole and seal with mud plaster. The holes may be plastered with mud prepared by adding Carbaryl 50 wp and soil in the ratio of 1:6 (*i.e.* 1 part of insecticide+6 parts of soil).

- ☆ Seriously affected trees cannot be saved with insecticide treatments and should be removed. Insecticides must be applied to kill adults as they are laying eggs on trunks and branches of trees. Spray Dimethoate 30 EC @ 100 ml/100 litres of water from June onwards for the control of adults.

4. Plum Aphid

Plum aphid *Hyalopterus pruni* Geoffroy (Aphididae: Hemiptera) is a highly destructive and polyphagous pest found on plum plants. The most obvious sign of these aphids on plum trees is the curled leaves they cause by their feeding. The pests are tiny and have shiny bodies that range from pale green to light yellow in color. The insect produces a high volume of honeydew, which is the excretion of the aphid. This in turn attracts ants that feed on the sweet liquid and causes a fungus to form sooty mold.

Host Plants

Plum, Peach, Apricot, Almond and Hawthorn.

General Appearance

The plum aphid is often found inside curled leaves. It is shiny and varies considerably in colour from green to brownish green or brownish yellow. The pest overwinters as eggs. Overwintering eggs are laid in crevices on twigs. They hatch just before bud break in the month of March to April. The nymphs develop into wingless females, which produce young ones without mating. The aphids feed on the undersides of leaves. As numbers increase, the leaves curl. After 2 or 3 generations, winged forms are produced in early June. They migrate from tree fruits to summer hosts, which include weeds, ornamental plants and vegetables. There, they produce several more generations up to August. In the fall, winged migrant aphids return to their fruit hosts and give birth to nymphs, which develop into egg-laying females. These females mate with males returning from summer host plants and produce eggs that overwinter. The pest may annually complete forty generations.

Nature of Damage

Colonies of this pest cause leaves to curl tightly. Often only one limb or a portion of a limb is infested early in the year. This aphid secretes a large amount of honeydew. Populations of this aphid can reduce both tree growth and fruit sugar content.

Management

- ☆ The best time to treat the pest is during the dormant or delayed dormant period. If aphids are a chronic problem in the orchard, a treatment in late fall/early dormancy is a very effective way to manage this pest. For good management of aphid secondary host plants should be removed from the field.
- ☆ There are many natural enemies that feed on plum aphid. However, fruit size may be reduced and curled leaves will not uncurl after aphids are suppressed. The recent introduction of *Anthocorids* has led to substantial levels of parasitism of this aphid. Important predators include brown lacewing, green lacewing, ladybird beetles and syrphid flies. Post bloom sprays should not be done in order to encourage the bio agents, which are effective in spring.
- ☆ Spray of narrow range horticultural mineral oil @ 2 per cent (2 litres/100 litres of water) during dormant season are organically acceptable method of controlling this pest as overwintering eggs can be killed at this stage.
- ☆ After petal fall spray Dimethoate 30 EC @ 100 ml or Chlorpyrifos 20 EC @ 100 ml or Phosalone 35 EC @ 140 ml or Methyl-*O*-demeton 25 EC @ 80ml/100 litres of water to control plum aphid.

5. Fruit Tree Leaf Roller Moth

Fruit tree leaf roller, *Archips argyrospila* Walker (Tortricidae: Lepidoptera) an early season pest that occurs on a very large number of fruit and ornamental trees Leaf roller caterpillars curl up the leaves of host plants and stick them together with silk webbing to make a shelter where they can feed on the leaves in safety. The young caterpillars feed on the inner surfaces of the leaves, but once the caterpillar approaches maturity it will eat through the leaf. This damage can be serious for host plants.

Host Plants

Plum, Almond, Apple, Apricot, Cherry, Pear, Quince, and Walnut.

General Appearance

The eggs are laid in oval batches of 20-100 on the twigs or bark of the fruit tree. The fruit tree leaf roller spends the winter in the egg stage on scaffold limbs and twigs and emerges in spring as larvae. Larvae feed on leaves for about 30 days and spin strands of silk and under ideal conditions disperse to other areas. The pest then pupates in a loose cocoon, which they form in a rolled leaf or similar shelter. Eight to eleven days later the adult emerges from the pupa. The moths live only for about a week, during which time they mate and lay eggs. They fly from May to June, depending on locality, and in any one area the flight usually lasts about three weeks. These moths lay eggs on twigs and branches, and the eggs will remain there until they hatch the next spring. Rolled leaves webbed together to form protective nests reveal the presence of leaf roller larvae. The pest has only one generation a year.

Nature of Damage

During bloom, larvae feed on leaves and buds. Later in the season they can feed on the surface of fruit, causing severe damage. Fruit often becomes infected with brown rot at feeding wounds.

Management

- ✰ Look for egg masses in late fall or early spring on the tree bark and in the joints of branches. Brush or scrape the eggs off the tree and dispose off them, far from susceptible plants.
- ✰ Release *Trichogramma* wasps at the first sign of adult moths. These minute, parasitic wasps lay their eggs in the eggs of other insects, reducing or eliminating the need for leaf roller control next season. Several native parasitic insects have provided adequate control of leaf rollers including species of Tachinid fly, Braconid, Ichneumonid, and Chalcid wasps.
- ✰ Dormant spray oil @ 2 per cent (2 litres/100 litres of water) gives effective control of overwintering eggs. Spray Dimethoate 30 EC @ 100 ml or Chlorpyrifos 20 EC @ 100ml/100 litres of water when larvae begin to emerge. Spray Phosalone 35 EC @ 140ml/100 litres of water during May to June to control the adults.

6. Plum Rust Mite

Plum rust mite *Aculus fockeui* Nalepa and Trouessart (Eriophyidae: Prostigmata) generally restrict their feeding to new foliage, causing these leaves to brown and roll upward longitudinally. Rust mites are tiny, microscopic mites that have two pairs of legs near the anterior end of the body. They are yellow to pinkish white to purplish in colour and wedge-shaped with the widest part of the body being just behind the head.

Host Plants

Plum, Cherry and Peach.

General Appearance

The life cycle is similar to that of apple rust mite.Plum rust mites overwinter as diapausing females in buds or in crevices of twigs and bark. Females leave their overwintering sites after the buds begin to expand in the spring. After feeding for a few days, they begin to lay eggs. A female lays 50-60 eggs, while males may deposit several hundred spermatophores. Eggs may take 4-15 days to hatch depending on the temperature. The immature stages require 2 to 18 days to complete. The life span of the adult is from 20 to 30 days. A complete generation requires 6 to 22 days. When environmental conditions provide the appropriate trigger, usually poor condition of foliage deutogynes are produced. This may be as early as mid summer. Males die in the fall. Females that are not inseminated produce only males, while those that are inseminated produce both male and female progeny. Overwintering mortality of females may limit the population the following spring. Dry weather also appears to reduce mite populations, perhaps through the effect of hardening the leaves. Females overwinter in dead or shrunken buds, moving to foliage as buds begin to open in spring. As many as 15 generations occur per year.

Nature of Damage

All species feed on leaves. Heavily infested leaves take on a silvery or bronze appearance, depending on the species. Severe infestations can interfere with photosynthesis; Mites feed on the surface of the leaf by piercing the epidermis with their mouthparts and sucking the fluids from the cells. On plums individual leaves may exhibit chlorotic fleck which is the occurrence of spots of abnormally yellow plant tissue up to 1-2mm in size. Mature plum foliage attacked by plum rust mites may be curled or dwarfed. The lower leaf surface has a bronze or silvery appearance. Feeding of mites on young foliage causes a toxemia, which is called yellow spot on peach and chlorotic fleck on plum. As the names imply, the symptoms are small yellow spots on the leaves, followed by shot holing in some cultivars.

Management

- ✰ Where this mite is troublesome, bearing and non-bearing trees can be sprayed during the dormant period before the buds swell in the spring with a lime sulphur solution. Predaceous mites of the genus *Typhlodromus* are by far the most effective enemies of the plum rust mite. In some seasons they almost eliminate it.
- ✰ Dormant oil spray @ 2 per cent (2 litres/100 litres of water) gives effective control of overwintering mites. Spray Propargite 57 EC @ 100 ml or Fenpyroximate 5 SC @ 100 ml or Dicofol 18.5 EC @ 108ml/100 litres of water.

The other pests that affect Plum are San Jose scale, *Quadraspidiotus perniciosus* (Comstock), Codling Moth, *Cydia pomonella* (Linnaeus), Root borer, *Dorysthenes hugelii* (Redtenbacher), Tent caterpillar, *Malacosoma indicum* (Walker) Blossom thrips *Frankliniella Schultzei* (Trybom) and Shot hole borer, *Scolytus nitidus* (Muller) are discussed under the pests of Apple.

Chapter 4

Insect Pests of Peach and their Management

Peach, *Prunus persica* Linnaeus (Batsch) is a deciduous tree or shrub in the family Rosacea grown for its edible fruit. The peach tree is relatively short with slender and supple branches. The leaves are alternately arranged, slender and pointed. The tree produces pink flowers, which have five petals and emerge in January and February. Peach may also be referred to as nectarine, the two fruits belonging to the same species, although nectarines have smooth skin, and are believed to have originated in China. Insect Pests are one of the constraints affecting production and quality of peach. Insect pests infesting peach have been recorded as sucking pests, beetles, and borers and are described as under:

1. Peach Leaf Curling Aphid

Peach leaf curling aphid *Brachycaudus helichrysi* Kaltenbach (Aphididae: Hemiptera) is a polyphagous pest native of Europe, and is now widely distributed across the world. Though a common pest of stone fruits, it is not usually considered a serious pest, except on nectarine. It is most prevalent both in plains and in the mountainous areas of India. It is more destructive on vegetable crops, which serve as summer host, and is a vector of several important diseases. Pear-shaped body of apterous female is yellow-green.

Host Plants

Plum, Wild plum, Peach, Apricot and Almond.

General Appearance

Overwintering takes place as fertilized eggs which are black, oval in shape and are located at bud base on young shoots of stone fruit trees. Hatching is observed at the end of March or April. The insects prefer lower side of leaves, forming large colonies and also feed on flowers. The aphids suck sap from leaf veins. Its secondary hosts are usually plants of the family compositae. Beginning of migration takes place at the end of May- June. Nymphal period lasts 6-11 days. Life span of apterous parthenogenetic females is about 24-49 days; their fecundity varies from 40 to 110 nymphs. The peach leaf curl aphid forms big colonies on both primary and secondary hosts. In August–September, the insects migrate to primary hosts, giving the sexual generation. Oviposition takes place from October, females lay eggs at bud base.The egg is shiny black and oval, similar to that of the apple aphid. The nymph is slender and pinkish in colour. As it develops, it becomes yellowish-green in colour. The wingless adult resembles the nymph but is larger. The winged adult has a black head and thorax and a yellowish-green abdomen with a large dark brown patch on the top of the abdomen. Overwintering eggs are laid in bud axils and bark crevices on twigs. They hatch just before bud break. The nymphs develop into wingless females, which produce young without mating. The aphids feed on the underside of the leaves. As the population increases, the leaves curl. After 2 or 3 generations, winged forms are produced, which migrate to summer hosts, where several generations are produced. In the fall, winged aphids return to peach and other stone fruits and give birth to nymphs. These develop into egg-laying females, which mate with males returning from summer host plants.

Nature of Damage

The species is oligophagous. It forms dense colonies on the underside of leaves of the primary host. Damaged leaves become yellow along veins while as diametrical wrinkles are formed on their surface. The leaves bend, roll, and dry up. Flowers also dry up, fruits fall down, and trees weaken. Fruit loss increases considerably.

Management

- ✰ Avoid planting aphid infested saplings. Secondary host plants should be removed from the field. Pruning of current years vegetative growth carrying eggs during December and January also reduce infestation.
- ✰ There are many natural enemies that feed on aphid however, fruit size may still be reduced and curled leaves will not uncurl after aphids are suppressed. The introduction of *Anthocorid bugs* has led to substantial levels of parasitism of this aphid. Important predators include: ladybird beetles, green lacewings, brown lacewings and syrphid flies.
- ✰ Sprays of Horticultural mineral oil @ 2 per cent (2 litres/100 litres of water) or neem oil is organically acceptable methods of controlling this pest. Spray Dimethoate 30 EC @ 100 ml or Phosalone @ 35 EC 140 ml or Methyl-o demeton 25 EC @ 80 ml/100 litres of water if curling of leaves is noticed.

2. Green Peach Aphid

Green peach aphid *Myzus persicae* Sulzer (Aphididae: Hemiptera) is the most significant aphid pest of peach trees, causing decreased growth, shriveling of leaves and the death of various tissues. It also acts as a vector for the transport of plant virus. The green peach aphid is found worldwide, although it is less tolerant to colder climates and overwinters through its eggs laid in trees of the genus *Prunus.* Plants that support aphids through the winter months include beet, brussels sprout, cabbage, kale, potato, and many winter weeds. These aphids can be transported long distances by wind and storms.

Host Plants

Peach, Plum, Apricot, vegetables *etc.*

General Appearance

Winter spent as eggs laid near buds of peach tree. In the spring, soon after the plant breaks dormancy and begins to grow, the eggs hatch and the nymphs feed on flowers, young foliage, and stems. After several generations, winged dispersants from overwintering *Prunus* spp. deposit nymphs on summer hosts. In cold climates, adults return to *Prunus* spp. in the autumn, where mating occurs, and eggs are deposited. All generations except the autumn generation culminating in egg production are parthenogenetic. Eggs are deposited on *Prunus* spp. trees. Mortality in the egg stage sometimes is quite high. Nymphs initially are greenish, but soon turn yellowish, greatly resembling viviparous adults. There are four instars in this aphid, with the duration of each averaging 2.0, 2.1, 2.3, and 2.0 days, respectively. Females gave birth to offspring 6 to 17 days after birth, with an average age of 10.8 days at first birth. The length of reproduction varies considerably, but averages 14.8 days. The average length of life cycle is about 10-20 days. The maximum number of generations annually is found to be 20. Around 8 generations may occur on *Prunus* in the spring, but as aphid densities increase winged forms are produced, which then disperse to summer hosts. Winged green peach aphids seemingly attempt to colonize nearly all plants available. They often deposit a few nymphs and then again take flight. This highly dispersive nature contributes significantly to their effectiveness as vectors of plant viruses. The offspring of the dispersants from the overwintering hosts are wingless, and each produces 30 to 80 nymphs. As aphid densities increase or plant condition deteriorates, winged forms are again produced to aid dispersal. In the autumn, in response to change in day length or temperature, winged male and female aphids are produced which disperse in search of *Prunus* spp. Females arrive first and give birth to wingless egg laying forms. Males are attracted to them by a pheromone, capable of mating with several females, and eggs are produced. The oviparous female deposits 4 to 13 eggs, usually in crevices in and near buds of *Prunus* spp. Parthenogenetic reproduction is favoured in many parts of the world where continuous production of crops provides suitable host plants throughout the year, or where weather allows survival on natural hosts.

Nature of Damage

Aphids feed on leaves; blossoms and fruit let resulting in fruit drop and sometimes, severe leaf curling. Green peach aphids can attain very high densities on young plant tissue, causing water stress, wilting, and reduced growth rate of the plant. Prolonged aphid infestation can cause appreciable reduction in yield of crops. Contamination of harvestable plant material with aphids, or with aphid honeydew, also causes loss. Blemishes to the plant tissue, usually in the form of yellow spots, may result from aphid feeding. The major damage caused by green peach aphid is through transmission of plant viruses. This aphid is considered to be the most important vector of plant viruses throughout the world. Nymphs and adults are equally capable of virus transmission but adults, by virtue of being so mobile, probably have greater opportunity for transmission.

Management

- ☆ Avoid planting aphid infested saplings. Secondary host plants should be removed from the field. Pruning of current years vegetative growth carrying eggs during December and January also reduce infestation.
- ☆ Yellow traps, are commonly used for population monitoring.
- ☆ There has been considerable success using parasitoids, the entomopathogenic fungus *Verticillium lecanii* (Zimmerman) and the predatory midge *Aphidoletes aphidimyza* (Rondani). Cultural manipulations may benefit predators and parasitoids. Bands placed around the trunks of peach trees can provide good harborage for predators that may suppress the aphids in the spring, thereby reducing the number dispersing to vegetables. Important predators include: ladybird beetles, green lacewings, brown lacewings and syrphid flies which reduces green peach aphid populations.
- ☆ Green peach aphid is quite responsive to alarm pheromone, which is normally produced when aphids are disturbed. Application of alarm pheromone has shown the potential to disrupt virus transmission, but this has yet to become an operational technology. A sex pheromone is also known from this aphid, but it functions only at short distances, and has not yet proved to be useful in aphid management.
- ☆ For overwintering eggs spray horticultural mineral oil @ 2 per cent (2 litres/100 litres of water) that will give adequate aphid control. Spray Chlorpyrifos 20 EC @ 100 ml or Dimethoate 30 EC @ 100ml/100 litres of water after leaves start to grow. Spray Methyl-*O*-demeton 25 EC @ 80ml/100 litres of water as soon as nymphs are observed.

3. Lecanium Scale

Lecanium scale, *Eulecanium* spp. Ferrris (Coccidae: Hemiptera) are sucking insects that insert their tiny, straw like mouthparts into bark, fruit, or leaves, mostly on trees and shrubs. Some scales can seriously damage their host, while other species do no apparent damage to plants even when scales are very abundant. The Lecanium scale insect is common on many types of trees and is prevalent on fruit trees. The appearance of Lecanium scale is dark crusty bumps, thick; white waxy bumps, or clusters of scaly bumps on stems and the underside of leaves. The bumps can often be scraped off with a dull knife, and the underside of the bumps is normally a softer material. When a tree is infected with Lecanium scale, it is very common to see a shiny, sticky substance on the leaves. This material is called honeydew and is an undigested sugar substance that is secreted by the scale insect. In certain climates, the honeydew becomes a prime-growing medium for black, sooty mould.

Host Plants

Peach, Apple, Pear, Plum, and Apricot.

General Appearance

Adult female scales and nymphs of most species are circular to oval, wingless, and lack a separate head or other easily recognizable body parts. Some scales change greatly in appearance as they grow, and some species have males and females that differ in shape, size, and colour. Adult males are rarely seen and are tiny, delicate, and white to yellow insects with one pair of wings and a pair of long antennae. Some scale species lack males and the females reproduce without mating. Scales hatch from an egg and typically develop through two nymphal instars before maturing into an adult. Each instar can change greatly as it ages so many scales appear to have more than two growth stages. At maturity, adult females produce eggs that are usually hidden under their bodies, although some species secrete their eggs externally under prominent cottony or waxy covers. Eggs hatch into tiny crawlers, which are yellow to orange in most species. Crawlers walk over the plant surface, are moved to other plants by wind, or are inadvertently transported by people or birds. Crawlers settle down and begin feeding within a day or two after emergence. Settled nymphs may spend their entire life in the same spot without moving as they mature into adults. Nymphs of some species can move slowly, such as soft scales that feed on deciduous hosts and move from foliage to bark in the fall before leaves drop. The pest is active from March to October. Lecanium scales spend the winter on twigs and branches in an immature or nymphal stage. Development resumes in the spring and mature females produce large numbers of eggs, which are protected by their soft waxy covering. Crawlers that hatch from these eggs move to leaves, settle, and feed on sap during the rest of the summer. They move back to twigs and branches prior to leaf drop and settle for the winter. As with many species of soft scale, these insects produce large volumes of a liquid waste called honeydew. A deposit of this sugar rich material gives leaves a shiny appearance. Limbs of heavily infested trees may be blackened by the growth of sooty mould fungus.

Nature of Damage

The signs of scale infestation are often noticed before one sees the insect itself. Most noticeable is the sticky, wet substance known as honeydew that is excreted in copious amount as the insects feed. Leaves and other surfaces may be lightly speckled with the shiny drops of honeydew, or they may be heavily coated with the sugary secretion and appear as if covered with shellac. Dieback of twigs and branches and premature leaf drop may result as heavily infested trees compete with scale insects for necessary moisture. There is presence of brown spherical grains on branches, yellow spots on leaves and drying of branches. The insect pierce the tissue and suck the sap and if abundant for 2-3 years, the infested branches may get killed and mould develops on exude of this pest.

Management

- ✰ Pruning and destruction of scale infested branches.
- ✰ Scale insects can be attacked by a variety of lady beetles, predatory mites, and small parasitic wasps. Conservation of these natural enemies should be done in the orchard by minimizing pesticide application.
- ✰ In early spring before leaves appear, horticultural mineral oil @ 2 per cent (2 litres in 100 litres of water) can be applied to the overwintering scales to suffocate them.
- ✰ Spray at bud burst stage any of the insecticides *i.e.* Chlorpyriphos 20 EC @ 100 ml/100 lit of water or Dimethoate 30 EC @ 100 ml/100 lit of water to control scale insects.
- ✰ If necessary repeat the spray 10 days after petal fall with Chlorpyriphos 20 EC @ 100 ml/100 litres of water or Dimethoate 30 EC @ 100 ml/100 litres of water or Malathion 50 EC @ 140 ml/100 litres of water.

4. Flat Headed Apple Tree Borer

The flat headed apple tree borer, *Chrysobothris femorata* Olivier (Buprestidae: Coleoptera) is a native pest of deciduous trees. It has a wide host range and can cause rapid decline of economically important hosts. When infestations are high, borer has been known to attack healthy trees as well as those stressed by drought, plant disease, mechanical injury and other environmental factors.

Host Plants

Apple, Pear, Peach, Apricot, Plum, Prune and Cherry.

General Appearance

Flat-headed apple tree borer completes one generation per year. Adult females lay eggs in the sun-exposed areas in bark crevices in late spring and throughout the summer months. Eggs hatch about seven days later; by the time grub chew through the bottom of the egg directly into the host tree, avoiding desiccation. Inside the host, grubs feed on the actively dividing cambium tissues, in addition to sapwood. Galleries, which are feeding paths made by grub, maintain optimal humidity for grub growth. Feeding activity continues even in cold winter months on sun-warmed portions of the trunk. When fully developed, grubs bore into the heartwood and form pupal chambers with entrances that are tightly plugged with frass. Pupation occurs in late spring to early summer and lasts 1-2 weeks, after which the adult emerges by cutting a distinctive D shaped exit hole in the bark.

Nature of Damage

The most serious injury to host plants is caused by grub feeding activity beneath the bark that damages the cambium layer and disrupts the flow of vital nutrients throughout the tree. A single grub is capable of girdling a young tree within one season. Evidence of grub activity can be found under bark of infested trees as sinuous feeding tunnels packed with frass. Portions of the trunk may show signs of infestation by noticeable oozing of sap. Trees that survive borer attacks are often left scarred and unproductive.

Management

- ☆ Borer's attacks newly transplanted trees, environmentally stressed, or have sustained bark damage. Localized damage of bark is opportune sites for attack. Drought stressed trees are especially vulnerable. Keep trees healthy, well fed and watered. Trees that have healthy vigour do not attract borers. Borers prefer stressed trees.
- ☆ To minimize borer attacks, use tree wraps, white trunk paint, or any type of tree growth shelter for newly planted trees. Pruning or other types of bark injuries should be avoided when the adults are present and active. Collect and destroy the damaged shoots and branches. Sanitation of fallen and standing dead wood and removal of any pruned materials is critical.
- ☆ Once the grubs have tunneled into the wood, insecticide sprays will not be effective. Insecticidal control for preventing further attacks is possible and should be applied over the bark early enough in the year to prevent successful adult egg laying. *Bacillus thuringiensis* (Balsamo) products have been found effective if applied when grubs are first noticed and before they tunnel into trunk. Some natural enemies like Gray field ant *Formica aerate* (Francoeur), Chalcid wasps *Copidosoma varicornis* (Nees) and *Hyperteles lividus* (Ashmead) are found feeding on the pest.
- ☆ Plug the holes with cotton soaked in Dichlorvos 76 EC @ 300ml/100 litres of water or Celphos @ 1 tablet/hole and plaster with mud.

5. Chafer Beetle

Adoretus simplex Sharp and *Holotrichia* spp. Hope (Scarabaeidae: Coleoptera) is a polyphagous pest, which feed on the foliage and fruits of about 300 plants, both wild and cultivated. Adults are shiny smooth beetles of various colour and size, which are adapted to feed on leaves of many different plant species. Adult beetles are medium sized, light reddish brown, and approximately 13–14 mm long. Grubs are white coloured and C shaped, with a yellow brown head and six jointed legs. Fully-grown grubs are 20–23 mm long. The pupae of the chafer resemble those of the other scarab beetles. Pupae are about 16 mm long. Eggs are shiny and oval, milky white when freshly laid, but later turning dull gray. The imago or adult beetle stage is quite short, lasting 1–2 weeks.

Host Plants

All fruit trees.

General Appearance

The adults come out of the ground in late spring and mate in large swarms, usually on low trees and shrubs. The beetles are most active on warm, clear nights when the temperature is above 19° C. They emerge at evening, mate through the night, and return to the soil by daybreak. Beetles may return to the trees to remate several times over the mating period. Female chafers lay 20-40 eggs over their lifespan. They are laid singly, 5–10 cm deep in moist soil, and take 2 weeks to hatch. The grubs hatch by late July. The grub population consists mainly of first instars in early to mid August, second instars by early September, and third instars by mid September to early October. In cold zones, the grubs feed until November, and then move deeper into the soil. In frost-free areas, the grub will feed all winter. Vigorous feeding occurs from March to May. In early June, the grubs again move into soil, up to 5 to 25 cm deep to form earthen cells and pupate. The pre pupal and pupal stages last 2 to 4 days and 2 weeks, respectively. By June, the new beetles start emerging.

Figure 11: Chafer Beetle Adult.

Nature of Damage

Adult beetle feeds on leaves, buds, blossoms and fruit lets. Eaten away leaves are perforated. The grubs feed on roots and may cause wilting of plant, stunted growth and finally death of plants. Beetles cluster at the top of the plants, so that upper canopy is usually defoliated more severely. In many fruit crops foliar damage is unlikely to affect yield. Once fruits become ripe, beetles may feed and aggregate on ripening fruit. The beetles feed during night and remain hidden during daytime. The pest remains active mostly from April to September.

Management

- Monitoring for chafer grubs is very necessary. April to May and September to October are the best times to monitor for the grubs because this is when they are the easiest to spot as they are at their largest stage. They swarm at dusk, sounding much like buzzing bees. When large populations are present, the evening swarms can be very obvious. Irrigation and mowing practices have also proved effective in suppressing and masking chafer problems. Frequent watering during chafer egg laying periods can give some protection.

- Autumn ploughing helps in the reduction of the pest by exposing the grubs to predators. Install a light trap near the orchard to collect and kill the beetles in kerosene oil.
- Some parasitic nematodes such as *Steinernema scarabaei* (Stock) and *Heterorhabditis bacteriophora* (Poinar) appear to be effective at reducing grub populations.
- Install Pheromone dispenser (3 to 4 dispensers/tree) on the host tree in the evening at the time of beetle emergence.
- Isolates of entomopathogenic fungi namely *Beauveria bassiana* (Balsamo) and *Metarrhizium anisopliae* (Metchnikoff) @ 108 spores/ml can be effective.
- Generally, 5 to 10 chafer grubs per square foot should warrant management tactics. If this pest spreads an insecticides should therefore be used judiciously. Spray Chlorpyrifos 20 EC or Dimethoate 30 EC @ 100 ml/100 litres of water during mass beetle emergence period. Apply Carbofuran 3 CG @ 100 g in the canopy area of the tree for management of grubs.

The other insect pests that affect Peach are: San Jose scale, *Quadraspidiotus perniciosus* (Comstock), Stem borer, *Aeolesthes sarta* (Solsky) Peach tree borer, *Sanninoidea exitiosa* (Say) Peach twig borer, *Anarsia lineatella* (Zeller), Root borer, *Dorysthenes hugelii* (Redtenbacher) Hairy caterpillar, *Lymantria obfuscate* (Walker), Bark beetle/Pin hole borer, *Scolytus nitidus* (Schedl), Indian gypsy moth, *Lymantria obfuscata* (Walker), Tent caterpillar, *Malacosoma indicum* (Walker) Blossom thrips *Frankliniella Schultzei* (Trybom) and Shot hole borer, *Scolytus nitidus* (Muller) are discussed under the insect pests of Apple and Plum.

Chapter 5

Insect Pests of Cherry and their Management

Cherry *Prunus avium* Linnaeus is a perennial tree in the family Rosacea grown for its fruit. Cherry occupies an important position among temperate fruits all over the world and is the season's first fruit to reach market. Cherry trees can live up to 60 years growing to a height of over 30 feet. Cherry may also be referred to as sweet cherry or mazzard. Cherry is native to Europe and Asia. Cherry fruit and trees are frequent target of several insect pests.

1. Black Cherry Aphid

Black cherry aphid, *Myzus cerasi* Fabricius (Aphididae: Hemiptera) commonly known as cherry aphid is found wherever cherries are grown throughout the world. It infests leaves and deposits honeydew on leaves and fruit, causing tight leaf curling. Heavy populations can deform shoots and stunting young trees.

Host Plants

Cherry, Apricot, Plum, Peach *etc.*

General Appearance

The pest overwinters as shiny black eggs on twigs and fruit spurs. Eggs hatch shortly before bloom and the aphids can go through a number of generations and may become very abundant in March or early April. The population decreases to a very low level on cherry trees during the summer months and primarily survives on mustard family weeds during this period. The body of apterous female is wide, pear like, shiny black from above and brown from below. Its length varies from 2 to 4 mm. Eggs are oval in shape. Eggs are located one by one at bud bases on young cherry trees. Nymphal period lasts 6 to 10 days. Aphids mature within 2 or 3 weeks of hatching and begin to reproduce asexually. Life span of apterous parthenogenetic female is 20 to 52 days; its fecundity varies from 60 to 120 nymphs. Parthenogenetic viviparous females of aphids appear in September. In October, the female lays 3 to 5 eggs on buds of cherry trees. Oviposition continues until leaf fall. In high damage zone, the largest number of aphids appears in May to June. In July to August the pest slows down its development, reducing fecundity because of hot temperatures and plant hardening. The most favorable conditions for apterous females are temperatures of 25 to 28°C and relative humidity of 70 to 80 per cent. The pest gives 8 to 10 generations during a year. In each generation some winged adults are formed, which migrate to plants of the mustard family. By early to midsummer few aphids remain on cherry. Succeeding generations occur on the summer host. In the fall, winged males and females fly back to the cherry where they mate and produce eggs that overwinter on the bark.

Nature of Damage

Black cherry aphids form dense colonies at the growing apices of cherry trees in spring. Initial damage is due to leaf curling. Continual feeding cause deformation of shoot growth and can also lead to the formation of pseudo galls. High populations of the black cherry aphid are mainly a problem on young trees where they cause curling and distortion of the leaves. The insects prefer lower side of upper leaves on young trees. On older trees, the pest feeds on lower leaves and young shoots. The Aphid also feeds on fruit stalks. The insects suck out sap of the veins; leaves fall in their growth, become twisted, then black and dry.

Management

- It is important to monitor cherry aphid during and shortly after bud break. Treatment is only feasible on trees small enough to be sprayed thoroughly. Lacewing larvae, ladybird beetles and syrphid flies are among the predators that help keep populations of this aphid under control.
- HMO @ 2 per cent (2 litres/100 litres of water) mixed with an insecticide (Ethion 50 EC @ 100 ml/100 litres of water applied to control other cherry pests usually control the cherry aphid. After the leaves curl it is difficult to achieve control through chemical and sprays may disrupt predators.
- Spray Dimethoate 30 EC or Chlorpyrifos 20 EC @ 100ml/100 litres of water at bud burst stage. Do not spray during the open blossom period due to the danger to bees and other pollinating insects.

2. Cherry Leafhopper

Cherry leafhopper, *Fieberiella florii* Stal (Cicadellidae: Hemiptera) is a minor pest of cherry. The pest is native to Europe.

Host Plants

Cherry, Plum, Peach, Almond, Pear and Apple.

General Appearance

Adult cherry leafhoppers are dark brown in colour but their shape and colour mimic the buds of their hosts. This leafhopper overwinters as eggs on ornamental hosts and deciduous fruit trees. Wingless nymphs hatch out and eventually develop into the winged adults in April. The larger nymphs and adults deposit white frass that resemble salt granules, which easily wash off. Most of the deposition occurs from April to June. Cherry is a preferred host for this pest. There are three periods of adult activity: mid April to May; July and September to October.

Nature of Damage

This leafhopper is a vector of diseases and can be responsible for severe outbreaks. Cherry leafhopper is an occasional pest. Damage is mainly due to nymphs producing honeydew. Honeydew accumulates on the fruit. The black sooty mold growing on the honeydew can be washed off, but the fruit may fail to colour.

Management

- ✫ Look for cherry leafhoppers from April to July. If easily found, apply an insecticide in July or August when nymphs are small.
- ✫ Spray Methyl-*O*-demeton 25 EC @ 80 ml or Dimethoate 30 EC @ 100 ml/100 litres of water respectively during July to August.

3. Cherry Fruit Fly

Cherry fruit fly *Rhagoletis indifferens* Curran (Drosophilidae: Diptera) is the most important pest of cherries. Once the skin of fruits becomes soft enough to penetrate, adult females insert eggs with their ovipositor, and maggot develop inside the fruits. The result is wormy fruit that is unmarketable. It is difficult to determine whether a fruit is infested until the maggot exits through a hole that it chews or the fruit is cut open to reveal the maggot inside.

Host Plants

Cherry.

General Appearance

Cherry fruit fly completes only one generation in a year. Cherry fruit flies spend roughly 10 months as pupae in the soil under the cherry trees, emerging as adults beginning in late May to October. Peak emergence often coincides with harvest. Adults live for 16 to 35 days, depending on temperatures. They feed on leaf deposits, such as honeydew and pollen. Adult females undergo a 7 to 10 day pre oviposition period before they are sexually mature. After mating, they lay eggs under the skin of the fruit. Females frequently feed on juices exuding from the puncture made during egg laying. Each female lay from 50 to 200 eggs in a 3-week period. The eggs hatch in 5 to 8 days, and the maggot burrow towards the pit of the fruit where they are unreachable by most pesticides. When fully developed, 10 to 21 days after hatching, they bore their way out of the cherries and drop to the ground. Within a few hours they burrow into the soil to pupate. The majority develops into adults the following season, though a few remain dormant for two years.

Nature of Damage

Adults do no damage the fruit. Maggots, which develop inside the cherries, make the fruit unmarketable. The maggots feed on the fruit pulp. Infested fruits may appear normal but when opened show broken brown areas where the maggots are feeding. In unsprayed trees a high percentage of fruit is likely to be attacked. The adult cherry fruit fly does not fly long distances so some unsprayed trees may remain un-infested for many years.

Management

- ✰ Monitor adult fly populations before fruit begin ripening and before flies begin laying eggs. Bucket style traps can be effectively used for monitoring the pest. Good field sanitation should be done to prevent further spread of flies. Ripe fruit should be picked frequently to minimize population buildup. All damaged fruit should be removed from the field and destroyed, either by burial or disposal in closed containers. Cherry fruit flies are weak fliers but spread easily through infested fruits or by wind. Therefore, all nearby sources of fruit should be managed to eliminate flies.
- ✰ Place methyl eugenol sex lure trap in the orchard @ 10 traps per ha to capture male flies. Use yellow sticky cards to detect when the first fruit fly emerges and apply the first spray within a week of the trap catch. Prepare bait with methyl eugenol 1 per cent mixed with Dichlorvos (76 EC) 0.1 per cent and put 10 ml of this mixture per trap and keep in different places.
- ✰ Insecticide sprays targeting the adult are the primary tactic for controlling this pest. Maggot of the fruit fly develops within the fruit where they are protected from most insecticides. Raking up soil below the tree and drenching with Chlorpyrifos 20 EC @ 2.5 ml/litre of water should be done to kill pupa.
- ✰ Spray Dimethoate 30 EC @ 100 ml or Malathion 50 EC @ 140 ml or Phosalone 35 EC @ 140 ml/100 litres of water respectively after petal fall to control fruit flies. Repeat spray if necessary after sweet varieties begin to develop colour.

4. Oriental Fruit Moth

The oriental fruit moth *Grapholita molesta* Busck (Tortricidae: Lepidoptera) originated in China and is now found in all fruit growing regions of the world.

Host Plants

Cherry, Peach, Quince, Apricot, Apple, Plum, and Pear.

General Appearance

The Oriental fruit moth is the primary pest of cherry in cherry growing areas. The larva is a smooth caterpillar with a shiny brown head and pinkish body. Eggs are laid singly on the surface of the fruits, in the first 1 to 2 weeks after petal fall. Egg deposition usually begins 2 to 5 days after the females emerge and continues for 7 to 10 days or longer. Each female lays 50 to 200 eggs. The eggs take 5 to 8 days to hatch. After hatching the neonate larva makes a small entry hole on the surface of the fruit thereby gets entry into the fruit. The larva matures in about 18 to 20 days varying with temperature, humidity and feeding conditions. Upon reaching maturity the larva stops feeding, leaves the fruit and undergoes pupation. The emergence of mature larva from the fruit can be detected by the presence of exit hole usually bordered with black margins and starts in the last week of May to June. After emergence the larva constructs a white silken cocoon above the ground and in a sheltered area under the piece of loose bark of the tree. The pupal period lasts for about 7 to 9 days. The adults survive for 11 to 13 days. The total survival period of one generation on cherry crop is 45 to 55 days. The number of generations per year varies from four to six.

Nature of Damage

The damaging stage of the pest is the larval stage, which feed on green fruits and cause premature ripening of the fruit. One larva damages a single fruit only. Infested fruits are filled with lot of frass pellets. The presence of larval frass inside the infested fruits usually provides them dark brown cloudy appearance. Rotten and soft skin of the fruit is indicative that the pest is feeding the fruit from inside. The larva completely feeds on the inner soft contents of the fruit leaving stone and upper layer of the fruit undisturbed. The stone of the infested cherry fruit are free and covered with fecal matter of the pest.

Management

- ✰ Discarding the infested fruits and good orchard sanitation is important in reducing oriental fruit moth populations. Remove and destroy dropped and culled fruits from the orchard. Unharvested and mummified fruit should be removed during pruning.
- ✰ Plant litter and other ground debris on the orchard floor should be kept to a minimum to reduce sites favourable for pupation and overwintering. Adequate suppression of the first brood may give control for the entire season.
- ✰ Oriental fruit moths are hard to control because once the insects have bored into the fruit it cannot be adequately controlled by pesticides. However, spraying the infested trees with Dimethoate 30 EC @ 100 ml or Chlorpyrifos 20 EC @ 100 ml/100 litres of water kill the surrounding night flying moths and prevent future worm damage.

The other insect pests that affect Cherry are: San Jose scale, *Quadraspidiotus perniciosus* (Comstock), Peach tree borer, *Synanthedon exitiosa* (Say), Chaffer beetle, *Adoretus Simplex* (Sharp), Stem borer, *Aeolesthes sarta* (Solsky), European red mite, *Panonychus ulmi* (Koch), Root borer, *Dorysthenes hugelii* (Redtenbacher), Blossom thrips, *Frankliniella Schultzei* (Trybom), Shot hole borer, *Scolytus nitidus* (Muller),

Green peach aphid, *Myzus persicae* (Sulzer), Lecanium Scale *Eulecanium* spp. (Ferris) and Flat headed borer, *Chrysobothris* spp. (Olivier) are discussed under the insect pests of Apple, Plum and Peach.

Chapter 6

Insect Pests of Almond and their Management

Almond, *Prunus amygdalus* Linnaeus including bitter almond, sweet almond, and flowering almond, is a small deciduous tree in the Rosacea family native to the Arabian Peninsula and western Asia, but now cultivated throughout the Mediterranean regions and temperate Asia. The tree has brown or gray bark and either an erect or weeping growth habit depending on the variety. Almond trees can reach height up to 13 to 33 feet and have a commercial lifespan of between 30 and 40 years. In India it is grown in Jammu and Kashmir. The major insect pests are as follows:

1. Almond Mealy Bug

Mealy bug *Drosicha dalbergiae* Stebbing (Margarodidae: Hemiptera) are soft-bodied, wingless insects that often appear as white cottony masses on the leaves, stems and fruit of plants. Adults are soft, oval distinctly segmented insects that are usually covered with a white or gray mealy wax. Small nymphs, called crawlers, are light yellow and free of wax. They are active early on, but move little once a suitable feeding site is found.

Host Plants

Almond.

General Appearance

Adult females deposit 300-600 eggs within an excreted, compact, waxy cottony appearing mass mostly found on the underside of leaves mostly confused with downy mildew. Egg laying is continued for about 2 weeks with the female dying shortly after all eggs are laid. Hatching occurs within 1-3 weeks and the small active yellow nymphs begin migrating over the plant in search of feeding sites on which to

Figure 12: Eggs of Mealy Bug under Magnification.

settle. As they feed, they secrete honeydew and a waxy coating begins to form over their bodies. Female nymphs pass through three stages with a generation taking as little as one month, depending on temperature. Male nymphs pass through five instars. They do not feed after the first two instars and exist solely to fertilize the females. The pest is found active from April to September. Overwintering nymphs become active in April to May synchronizing with the resumption of activity of the almond plant after dormancy. The average number of adults per colony varies from 5.5 to 8.5 and average number of nymphs per colony varies from 8 to 10.5.

Nature of Damage

They feed by inserting long sucking mouthparts, called stylets, into plants and drawing sap out of the tissue. Damage is not often significant at low pest

Figure 13: Egg Mass of *Drosicha Dalbergiae* in White Cottony Silky Mass.

levels. However, at higher numbers they can cause leaf yellowing and curling as the plant weakens. Feeding is usually accompanied by honeydew, which makes the plant sticky and encourages the growth of sooty mould. The insect is covered with white cottony mass below the collar of the plants. Old and weak plantation is most susceptible to attack and exhibit nutrient deficiency and sickly appearance, which result in poor yield.

Management

- ✫ Prune out light infestations or dab insects with a cloth dipped in absolute alcohol. Do not over water or over fertilize because mealy bugs are attracted to plants with high nitrogen levels and soft growth.
- ✫ Commercially available beneficial insects, such as green lacewing *Crysoperla carnea* (Stephens) and the Mealy bug destroyer *Cryptolaemus montrouzieri* (Mulsant), are important natural predators of this pest.
- ✫ Insecticidal soap contains potassium salts of fatty acids, which penetrates and damages the outer shell of soft-bodied insect pests, causing dehydration and death within hours.
- ✫ Soil insecticide Carbaryl 10 per cent dust @ 85gm/plant should be applied followed by Chlorpyrifos (10G) @ 83gm/plant and Quinalphos (5 per cent dust) @ 167gm/plant.
- ✫ Foliar spray of insecticide like Methyl-*O*-demeton 25 EC @ 80 ml, Dimethoate 30 EC @ 100 ml and Chlorpyrifos 20 EC @ 100 ml/100 litres of water can be sprayed to almond trees during the month of May.

2. Almond Leaf Roller

Fruit tree leaf roller *Archips argyrospila* Walker and Oblique banded leaf roller *Choristoneura rosaceana* Harris (Tortricidae: Lepidoptera) are the most common leaf roller pests throughout the almond growing areas. It is native to North America but has been accidentally introduced into other parts of the world.

Host Plants

Almond, Apple, Apricot, Cherry, Pear, Plum, Prune, Quince, and Walnut.

General Appearance

Fruit tree leaf rollers overwinter in the egg stage on twigs. Eggs hatch into tiny larvae from March to as late as mid May. Larvae are dark green with black heads and are about 1 inch long. When fully grown they are difficult to distinguish from oblique banded leaf roller. Larvae feed on leaves for about 30 days then pupate in a loose cocoon, which they form in a rolled leaf or similar shelter. Eight to 11 days later the adult emerges from the pupa in June or July and deposit overwintering eggs. The moths live only about a week, during which time they mate and lay eggs. They fly from May to June, depending on locality, and in any one area the flight usually lasts about three weeks. These moths lay eggs on twigs and branches, and the eggs will remain there until they hatch the next spring. There is only one generation in fruit tree leaf roller.

Oblique banded leaf roller overwinter primarily as second or third stage larvae in protected places in trees and under bud scales. The overwintered larvae become active as the buds begin to open. They begin to feed by tying together a number of leaves with silk. They first feed on water sprouts and then move throughout the tree. Those feeding on developing flower buds do so before bloom and continue to consume floral parts throughout the blossom period. This is when they cause the most damage to the almond crop. After petal fall, these larvae continue to feed on developing fruit. Emerging larvae are greenish yellow caterpillars, usually with black heads but sometimes with lighter coloured heads. Pupation occurs within these sheltered areas and the adult moths generally appear during late May and early June. Eggs are laid in flattened, overlapping masses of up to 300 on the upper surface of leaves. The second generation of these leaf rollers, which occurs in June or July, is more likely to occur on trees, causing damage later in the season than the fruit tree leaf roller. Adults appear bell shaped when at rest and have dark brown bands running at oblique angles across their wings. The wings are mottled with gold and white flecks. Oblique-banded leaf rollers occur on a wide range of plants. There are two to three generations in oblique-banded leaf roller.

Nature of Damage

Leaf rollers are occasional pests of almonds. The primary damage occurs early in the season when larvae of the overwintered generation feed on developing nuts and hollow them out. Many of the damaged nuts are lost in the June drop presumably reducing yield. The summer generation of the oblique banded leaf roller ties leaves and nuts together and feeds on the hulls.

Management

- ✰ A number of insects eat leaf rollers including certain Tachinid flies and Ichneumonid wasps, which parasitize the larvae. After consuming the leaf roller larvae, the Braconid wasp forms a white cocoon next to the shriveled up worm inside its nest. A white cocoon is an indication that the parasite is present and might provide control. Lacewing larvae, assassin bugs, and certain beetles also are common predators. Birds sometimes feed on the larvae and pupae. These natural enemies often help to keep leaf rollers at low, non-damaging levels, but even if natural enemies are present, large outbreaks of leaf rollers occasionally occur.
- ✰ The microbial insecticide *Bacillus thuringiensis* (Balsamo) is effective against the larval stages of leaf rollers. Leaf rollers stop eating within hours after feeding on a sprayed leaf and die several days later. Thorough spray coverage of the tree is required for control. *Bacillus thuringiensis* is most effective on leaf roller larvae when they are small (less than ½ inches long) and usually requires more than one application.
- ✰ Spray the orchard with Chlorpyrifos 20 EC @ 100 ml or Quinalphos 25 EC @ 100 ml/100 litres of water.

2. Popular Bug

Popular bug *Monosteira unicostata* Mulsant and Rey (Tingidae: Hemiptera) is a common pest of fruit trees, also called false tiger of fruit trees. Poplar bug is one of the most important pests of Almond.

Host Plants

Prunus, Malus, Populous, Salix *etc.*

General Appearance

The pest is active from May to August. Popular bug overwinters as adults inside the bark crevices of trunk and large branches of host trees, or beneath the dried leaves or debris around tree bases. Egg laying starts from late April or early May and females lay eggs inside leaf lamina. The average number of eggs laid per overwintering female is 33 to 41 and the average oviposition period is 11 to 33 days. The insect has five instars. The nymphs often form the colonies beneath the leaves and suck sap of leaves. The first generation adult appears from early June and egg laying starts from mid June. The second and third generation adults are found in early and late July respectively and the fourth generation adults are found in mid or late August. The insect has 3 to 4 generations per year. The life span of overwintering adults of third and fourth generation including winter diapause is about 8 months.

Nature of Damage

The pest sucks the sap from the plants resulting in yellowing of plants and premature fall, with negative consequences for the following year. The pest causes drying of leaves, which reduce the production of the crop. The female causes damage, which inject their eggs into the leaf parenchyma and by the adult and nymphs, which cluster around the principal veins to suck the sap.

Management

- ✰ When the infestation is observed in 10 per cent of the leaves the treatments of pesticides gives good results. The chemical should be applied in June before the future fall of leaves. Repeat the treatment after 15 days to cover the entire period of hatching eggs.
- ✰ Spray Dimethoate 30 EC @ 100 ml/100 litres of water to suppress overwintering adults. Spray Methyl-o-Demeton 25 EC @ 80 ml/100 litres of water in May to June when nymphs start appearing.

4. Almond Weevil

Almond weevil *Myllocerus laetivirens* Marshall (Curculionidae: Coleoptera) is less than 6 mm in size that feed on the almond kernel. Adults are small, 3 to 4 mm long, green with yellow tinge. Scales are present on the body into yellowish greeen or mettalic green in colour.

Host Plants

Almond, Plum, Apricot, Mulberry, Apple, Pomegranate, Strawberry and Pear.

General Appearance

The first adult begins to appear from the month of April to May and lay eggs from the end of July to September in the batches of 40 to 50 each. The incubation period is of 4 to 5 days and the grub burrow deep in the soil up to 200 to 300 mm. They feed on roots of the plants. The grubs have five instars, which last for 2, 2, 3, 3, and 4 days respectively. When full grown they come up on the soil surface to pupate in the upper 25 mm of the soil. The pest passes winter in the pupal stage in the soil. At the start of summer, the adults move into various shelters like cracks in the bark of almond or other nearby trees. The adult measures 3 to 4 mm. The pest has one generation in a year.

Nature of Damage

The weevils congregate on the ventral surface on leaves, nibble irregular holes and gradually eat away the entire leaf lamina, leaving only the midribs. The tender leaves are eaten away and later on older leaves are skeletonized. The pest causes the maximum damage to foliage during rainy season.

Management

- ✰ Cultural practices may be of value. Frequent hoeing and intercultural operations disturb and kill the grubs of the weevil. Collect attacked and fallen fruits and destroy them.
- ✰ Apparently several insecticides are effective against adult weevils however; the chemical treatment may be of little or no economic value because of the limited period of vulnerability. The grub will probably be protected while feeding under the ground. However, Carbofuran 3 CG @ 700 gm/tree appears to be effective against pupae in the soil. Spray Quinalphos 25 EC @ 100 ml or Chlorpyrifos 20 EC @ 100 ml/100 litres of water.

The other insect pests that affect Almond are: San Jose scale, *Quadraspidiotus perniciosus* (Comstock), Oriental fruit moth, *Grapholita molesta* (Busck), Stem borer, *Aeolesthes sarta* (Solsky), Almond tree borer/American plum borer, *Euzophera semifuneralis* (Walker), Tree borer, *Sanninoidea exitiosa* (Say), Peach twig borer, *Anarsia lineatella* (Zeller), European red mite, *Panonychus ulmi* (Koch), Root borer, *Dorysthenes hugelii* (Redtenbacher), Indian gypsy moth, *Lymantria obfuscata* (Walker), Tent caterpillar, *Malacosoma indicum* (Walker), Blossom thrips, *Frankliniella Schultzei* (Trybom), Shot hole borer, *Scolytus nitidus* (Muller), Peach leaf curling aphid, *Brachycaudus helichrysi* (Kaltenbach), Chaffer beetle, *Adoretus simplex* (Sharp), *Holotrichia* spp (Hope) and June beetle, *Polyphaga decemlineata* (Say) are discussed under the insect pests of Apple, Plum Peach and Apricot.

Chapter 7

Insect Pests of Apricot and their Management

Apricot, *Prunus armeniaca* Linnaeus is a deciduous tree in the family Rosaceae grown for its edible fruit. Apricot trees can reach 26 to 39 feet and can live anywhere between 20 and 40 years depending on variety and growth conditions. The origin of apricot is disputed. It was known in Armenia during ancient times, and has been cultivated there for so long that it is often thought to have originated there. There are many types of insect pests on apricot trees, but most can be controlled without using insecticides. The insect pests affecting apricot are as under:

1. Apricot Chalcid

Apricot chalcid *Eurytoma samsonowi* Vassiliev (Eurytomidae: Hymenoptera) is the major pest of apricot in all apricot growing areas of the world.

Host Plants

Apricot.

General Appearance

The insect pest attacks the fruits immediately during fruit set in April to May and deposits the eggs inside the young fruits. The larvae feed on the contents of the kernels and complete development during fruit maturation stage and ultimately the fruit drops. The fallen fruits show symptoms of flesh shrinkage and are also of reduced size. The larvae are translucent white and feed during April to June and thereafter remain quiescent from July to February. They then pupate inside kernels and with the onset of warm weather adult chalcid wasps appear after piercing through the stones and oviposit after mating. There is only one generation in a year.

Nature of Damage

The grubs bore into the kernels and feed in the inner contents leaving the papery coat intact. As a result fruit development is arrested and the fruits fall prematurely with larvae still feeding within the fruits. Larvae, which are translucent white, feed during April and May and remain quiescent from June to February.

Management

- ✰ Keep the trees as free as possible from mechanical wounds, winter injury, crotch separation and cankers. All the infested and fallen fruits should be collected and destroyed to control this pest.
- ✰ Spray horticultural mineral oil @ 2 per cent (2 litre/100 litres of water) during dormancy when the temperature is above 4°C to control the pest to some extent.
- ✰ Pre bloom spraying against aphids and mites will also provide control against chalcid. However all fallen fruits that may contain the larvae be collected and burnt.

The other insect pests that affect Apricot are: Branch and twig borer, *Melalgus confertus* (LeConte), San Jose scale, *Quadraspidiotus perniciosus* (Comstock), Mealy plum aphid, *Hyalopterus pruni* (Geoffroy), Fruit tree leaf roller, *Archips argyrospila* (Walker), Chaffer beetle, *Adoretus Simplex* (Sharp), Stem borer, *Aeolesthes sarta* (Solsky), Peach twig borer, *Anarsia lineatella* (Zeller), European red mite, *Panonychus ulmi* (Koch), Root borer, *Dorysthenes hugelii* (Redtenbacher) Indian gypsy moth, *Lymantria obfuscata* (Walker), Tent caterpillar, *Malacosoma indicum* (Walker) Blossom thrips, *Frankliniella Schultzei* (Trybom), Shot hole borer, *Scolytus nitidus* (Muller) and Peach leaf curling aphid, *Brachycaudus helichrysi* (Kaltenbach) are discussed under the insect pests of Apple, Pear, Plum and Peach.

Chapter 8

Insect Pests of Walnut and their Management

Walnut *Juglans regia* Linnaeus is believed to have originated from Iran. In India, only the state of Jammu and Kashmir is considered as hub for walnut production being beneficial of its medicinal, economical and nutritional values. However, during the recent past it has spread to other states like Himachal Pradesh, Uttarakhand and Arunachal Pradesh. The walnut fruit industry is the backbone of the economy of Jammu and Kashmir as its cultivation is directly connected to economic prosperity of the people. Walnut (*Juglans regia*) is included in Food and Agricultural Organization (FAO) list of priority plants because of its nutritive value. Several insect pests depending upon season and location infest walnut orchards. The damage results in defoliation, fruit drop and finally yield loss.

1. Walnut Blue Butterfly

Walnut blue butterfly *Chaetoprocta odata* Hewitson (Lycaenidae: Lepidoptera) is considered as one of the most serious lepidopteran insect pest infesting walnut trees.

Host Plants

Walnut.

General Appearance

The pest is active from March to July. The female adult after hovering over host plant twigs are seen depositing her egg mass on selected part of host plant twigs. The egg mass is very difficult to identify as it camouflage with host twigs. The egg mass contains about 14 to 22 eggs. A group of rounded creamy white eggs are sometimes observed on bark surface of the walnut twig. On approach of summer season, the eggs hatch and neonate larvae emerge that coincides with the sprouting of buds. The first instar larva emerges from the egg case by gnawing away an irregular hole with the help of its mandibles and wriggles out by enlarging the slit through the peristaltic movements. Emergence leads in an irregular split anterio posteriorly at least half way down the egg capsule length. The slug like larvae is greenish in colour and flattened in shape. The neonate larvae after coming out from the eggshell start feeding on buds before it starts actual feeding on the foliage. The neonate caterpillars are cylindrical in shape and light green in color. Pupation of the insect take place on walnut leaves itself during late May and adults emerge during early June to end of June which are blue coloured moths flying actively at dusk around the walnut trees. Walnut blue butterfly is a monophagous pest having only one generation in a year.

Nature of Damage

The pest causes defoliation to walnut trees. The caterpillars in heavy infestation eat up leaves and emerging buds of the plants.

Management

- ☆ All the infested nuts should be collected and buried deep into the soil. Sanitation of orchard is necessary to control the pest.
- ☆ When the infestation is severe the pest should be sprayed with Dimethoate 30 EC @ 100 ml or Phosalone 35 EC @ 140 ml in 100 litres of water in early April.

2. Walnut Tortrix Moth

Tortrix moth *Archips argyrospilus* Walker (Tortricidae: Lepidoptera) native of Europe and parts of Asia is a pest of deciduous trees and shrubs. The pest is a leaf feeder and incidental feeding on fruit is caused when the leaves are webbed to the fruit.

Host Plants

Walnut, Apricot, Plum, Peach, Apple, Chestnut, Hazelnut *etc.*

General Appearance

The pest is active from July to September. Female lays eggs on the fruit or on the stalk of the fruit and are covered with grey secretion that turns white upon aging. The incubation period is two to three weeks. Larvae have seven instars. First instar larvae bore into the buds of their host plant. They feed on the leaves, flowers, buds or fruits of the host plant. Later instars roll or tie leaves together or to fruit and partially emerge from the shelter to feed. The infestation is manifested by the presence of a hole generally near the stalk end where the two fruits come in contact with each other. The black excreta along with frass come out from the holes. Larvae may cause further damage to the leaves before they hibernate. Larvae usually overwinter in forked branches or bark indentations, where they emerge in late spring or early summer to begin another cycle. Pupation takes place in late May and early June into cocoons inside the rolled leaf, emerge in late summer and early autumn. The adult moths appear after 10 to 12 days in mid June and are found until mid August. The adult flight period lasts approximately 3 weeks. There is one generation in a year but sometimes if temperature is favourable two generations are also completed.

Nature of Damage

The damage in buds and blossoms can be seen as small entry holes, chewed petals and flower parts. Petals are webbed together, often remaining attached through petal fall; inner flower parts are eaten. Leaves are also chewed, rolled and tied together with silk. The caterpillar bore the fruit hull and makes tunnels in the hull and thereby turns the hull black in colour. Kernel inside the nut also turn black thereby rendering unfit for the market. Because of rotting, bacterial infection occurs and black juices are coming out from the fruit. The damaged nuts are dry and collapsed with large slot like holes and russeted scars are also found on fruits.

Management

- ✰ All the infested fruits and leaves should be collected and buried deep into the soil so as to have reduced population next year. Complete sanitation of orchard is necessary.
- ✰ The hymenopteran species belonging to three families (Ichneumonidae, Braconidae and Chalcididae) and one parasitic dipteran species (Tachinidae) parasitize the larvae and pupae of the pest.
- ✰ When the infestation is severe the tree should be sprayed with Dimethoate 30 EC or Chlorpyrifos 20 EC @ 100ml/100 litres of water during the onset of May.

3. Grey Weevil

Grey weevil, *Myllocerus* spp. Marshall (Curculionidae: Coleoptera) causes damage to the foliage and possibly root systems Adults feeding is most noticeable when plants are producing new foliage. About 336 species of genus *Mylocerus* are known from the world. So far 89 species are reported from India, infesting forest trees and horticultural crops.

Host Plants

Walnut, Peach, Plum, Strawberry, Loquat, fig *etc.*

General Appearance

The pest is active from April to September. Female *Mylocerus* spp. lay up to 360 eggs over a 24-day period singly in a soft organic matter on the ground. Eggs are less than 0.5 mm, white or cream coloured at first, and then gradually turn brown when they are close to hatching. The grubs emerge in 3 to 5 days. They burrow into the soil where they feed on plant roots for approximately one to two months. The grubs pupate in the soil for approximately one week. The adult have been found to feed on foliage of walnut leaves by biting holes in the lamina or by peripheral feeding on margins and later show a burnt blackish appearance.

Nature of Damage

When adult weevils feed on leaves, they feed inward from the leaf margins, causing the typical leaf notching. There are some instances where the leaf material is almost completely defoliated, where the weevil has fed along the leaf veins. The adults prefer new plant growth. Intense feeding by numerous weevils may cause plant decline or stunting. Young seedlings may not survive a large amount of feeding damage. With healthy plants, however, the feeding damage may be considered cosmetic if the plant recovers.

Management

- ☆ Collect and destroy the adults when there is infestation. Fallen fruits should be collected at weekly interval till fruit harvest. Plough the orchard after harvest to expose hibernating adults. Destroy all left over nuts in the orchard and also in the processing industries.
- ☆ Spray Chlorpyrifos 20 EC @ 100 ml/100 litres of water during April to May controls the pest population built up.

4. Walnut Weevils

There are two weevils that attack the walnut fruit:

a. Butternut Curculio

Conotrachelus juglandis Le Conte (Curculionidae: Coleoptera) commonly known, as butternut curculio is a snout weevil, about 6 mm long, brownish grey in color, with a broad whitish band across the wing covers. The grubs are plump, with a brown head. Injury to the flowering shoots in early spring and injury to the developing nuts during the summer usually causes premature abortion and drop of fruit. The entire nut crop can be lost due to curculio injury in severe cases. In orchards, grubs that feed on new stems can kill new branches of the previous season's growth.

Host Plants

Walnut.

General Appearance

The pest is active from April to September. The weevil is univoltine and adult live for a year or longer. The weevil is black in colour and prefers superior thin-shelled varieties to less valuable thick-shelled varieties. The females oviposit from mid spring to late summer, so there is considerable overlap of the life stages. The female weevil oviposits in niches or cavity, which she made by chewing in the young shoots. The grubs emerge, start feeding and live for 4 to 6 weeks. Fully developed grubs, exit through the tunnel opening and drop to the ground. They burrow into the soil as deep as 3 inches and construct a small earthen pupal cell. Pupation occurs 7 to 10 days after grub enters the soil and lasts for 15 to 20 days. Adults appear from July to September. Mostly feeding by adult occurs during night and morning hours. They feed on the soft tissue of the shoot tips and on the leaves. As the leaves begin to shed and cold weather arrives the adult drop to the ground, and burrow into the soil beneath the trees canopy.

Nature of Damage

The grub bore inside the walnut fruit feed and turns the kernel into a black powdery mass that cause the great injury. In the spring adults make large punctures in the leaf, stems, and young shoots sometimes causing them to wilt and die. In some instances the new shoots are killed entirely due to feeding by adult.

b. Walnut Weevil

Alcidodes porrectirostris Marshall (Curculionidae: Coleoptera) was first reported from Pakistan on the walnut plantation. The pest injures the nuts and shoots of walnut. Adults feed on the buds and flowers but the grub feed inside the fruit and are responsible for causing premature dropping.

Host Plants

Walnut.

General Appearance

Adults overwinter within the bark of trees and emerge in early summer to colonize on host trees. Adult lays eggs in young green fruits of walnut during first and second week of May. First and second instar grubs feed on the green skin and flesh and the third instar bore through the hull and reach the kernel of the fruit. At this time most of the infested fruits drop to the ground. Larval period lasts for about 45 days. Grubs start pupation inside the fallen fruits by early July and the pupal period last for about 14 days. Adults start emerging while still inside the fruits in the last week of July and early August. Fruit fall due to weevil infestation starts from middle of May and continue to the end of July, most fruits fall during the third week of June. All the developmental stages are found inside the fallen fruits.

Nature of Damage

The pest is active from May to September. Weevil damages leaves and fruits during June to July. Fruits turn into black mass and are totally damaged. Jet-black weevil start appearing in early May, deposit eggs after mating on the green nutlets, which after hatching into grubs bore the nuts and cause the damage.

Management

- ✰ Monitor pest activity in spring, when shoot growth begins. Collect and destroy fallen fruits. Sanitation should be observed in the orchard. Watch for signs of brownish to black adults, black feeding scars on new shoots and crescent egg laying scars on the nuts. Pruning should be disposed of promptly. Large wood should be ground up, or removed and burned.
- ✰ When the infestation is severe the tree should be sprayed with Dimethoate 30 EC @ 100ml or Quinalphos 25 EC @ 100 ml or Chlorpyrifos 20 EC @ 100 ml in 100 litres of water in the month of May.

5. Walnut Husk Fly

Rhagoletis completa Cresson (Tephritidae: Diptera) was first reported in California in 1926. It is a mid to late season pest of walnut which is the suitable hosts of husk fly. It occasionally attacks peaches grown close to walnuts. The walnut husk fly remains on a single walnut tree or group of trees as long as there are plenty of nuts. A yellow spot just below the area where the two wings are attached and a dark triangular band at the tip of wings distinguish the husk fly from other flies to be found in orchards. The adult is about the size of a housefly and is very colourful.

Host Plants

Walnut, Peach, Cherry.

General Appearance

The pest is active from July to September. The pest overwinter as pupae in the soil, emerge as adult from early July. Peak emergence is found in mid of August. After mating, the female lay its eggs in groups below the surface of husk. Eggs are laid in groups of about 15 in a small cache beneath the fruit skin. A freshly laid egg is pearly white. Shortly before hatching, the mouth hooks of the young larva can be seen. The eggs hatch into white yellowish maggots with black mouthparts in 5 to 7 days depending on the temperature and the young ones begin to feed on husk. The maggots develop in 3 to 5 weeks and later drop to the ground and enter the soil and pupate. The pupa is straw colored with conspicuous dark brown anterior spiracles. It is of the size and shape of a large grain of wheat. The adult is about 6 mm long, slightly smaller than a housefly. It has large, iridescent, greenish eyes, a tawny body colour and banded wings. It has a yellow spot on its back. Its conspicuous wing markings make it easy to identify. The female is larger than the male and its abdomen is more pointed. There is only one generation a year.

Nature of Damage

The primary damage from the husk fly is nutshell staining, which is a problem in commercial orchards where nuts are grown. Feeding by the husk fly maggots also causes the damaged husk to stick to shell, making them difficult to remove. An early season husk fly infestation can result in shriveled, moldy kernels.

Management

- ✰ Remove and dispose off damaged nuts as soon as possible. Yellow sticky traps can also be placed under the tree from July through August to prevent the maggots from entering the soil to pupate and to detect emergence. Nuts should also be examined to see the number of egg punctures and therefore the time of first application of chemical is decided.
- ✰ The pest can be controlled by the application of Chlorpyrifos 20 EC @ 100 ml, Dimethoate 30 EC @ 100 ml and Methyl-*O*-demeton 25 EC @ 80 ml in 100 litres of water. Apply again after 10 days if the husk fly was a problem the previous year. A third application may be needed 3 to 4 weeks later if flies continue to be caught in traps.

6. Walnut Aphid

Walnut green aphid *Chromaphis juglandicola* Kaltenbach (Aphididae: Hemiptera) occurs as a pest on the walnut tree causing severe damage. It is yellowish in colour and is usually found feeding underside of leaves resulting in reduction of yield.The infestation of aphids can cause withering of leaves and growing of sooty mould resulting in nut damage.

Host Plants

Walnut.

General Appearance

The pest is active from April to October. It is a medium pest of walnut. The pest is most serious in spring and early summers when rapidly growing nuts are particularly susceptible to stress. In the fall wingless females migrate to the trunks of walnut trees, where they deposit orange coloured eggs in cracks. Eggs overwinter and hatch during bud burst. After maturity they reproduce parthenogenetically. After hatching in spring the nymphs migrate to the upper surface of tender leaves. Winged and wing less females appear in the fall in the month of September to October. Depending upon the temperature the population increases. Aphids pass through many generations a year. In the fall the wingless females as a result of mating with winged males lay overwintering eggs.

Nature of Damage

Aphid feeding can reduce tree vigor and nut size, yield, and quality. As aphid excrete honeydew, sooty mould growing on the honeydew turns the husk surface black, and increases the chance for sunburn on exposed nuts. High populations of aphid may lead to leaf drop, exposing more nuts to sunburn, which darkens or shrivels the kernels. The aphids feed on the sap of leaves, causing drying of leaves. The pest also causes stunted growth in nursery plantations.

Management

- Monitor the pest before bud break as the aphids are small at that stage by taking 5 first sub terminal leaflets from 10 trees throughout shoot and nut growth period.
- There are many natural enemies affecting walnut aphid including ladybird beetles, syrphid fly larvae and green lacewing. The parasitic wasp, *Trioxys pallidus* (Haliday) is a good biocontrol agent. Only few orchards require treatment for walnut aphid, except when the parasite is disrupted by chemical treatments for other pests.
- Walnut aphid populations will often increase rapidly if chemicals are applied that interfere with biological control or if the hyper parasites are not controlled. Consider treatments for walnut aphid if the average number of aphids found on the underside of sub terminal leaflets is over 15 per leaflet.

- Horticultural mineral oil (HMO) @ 2 per cent (2 litres/100 litres of water) is effective against the pest. Summer spray of HMO @ 0.75 per cent (750 ml/100 litre of water) is also effective.
- When the infestation is severe the tree should be sprayed with Dimethoate 30 EC @ 100ml or Chlorpyrifos 20 EC @ 100 ml or Methyl-*O*-demeton 25 EC @ 80 ml per 100 litres of water during May to early June.

7. Dusky Veined Aphid

Dusky veined aphid *Callipterus juglandis* Goetze (Aphididae: Homoptera) is one of the potential walnut pests. The life cycle of dusky veined aphid is similar to walnut green aphid. They are bigger in size than walnut aphid. The pest prefers to feed in rows along the mid veins of upper surface of leaves. Females have a dusky marking while as nymphs have a dark banded spots on the back. These spots are much less pronounced or absent on the nymphs of walnut aphid.

Host Plants

Walnut.

General Appearance

The pest is active from April to October. The infestation on leaves is higher in the month of June and then it reduces gradually. Other appearance is same as walnut Aphid.

Nature of Damage

The pest feeds on the upper side of leaves in rows, sucks the sap and cause premature leaf drop, reduced nut size and yield apart from production of large amount of honeydew which invite growth of sooty mould giving a blackish appearance in the leaf and husk of developing nuts.

Management

Treatments should be considered for dusky veined aphid whenever an average of 10 per cent of the sub terminal leaflets have dusky veined colonies of six or more feeding on their upper surface along the mid vein. Management of the dusky veined aphid is same as that of walnut Aphid.

8. Walnut Scale

Walnut scale *Quadraspidiotus juglandis* Comstock (Diaspididae: Hemiptera) is an apparently native armored scale insect that occurs throughout the walnut growing areas of the world. It attacks a wide variety of woody plants ranging from deciduous trees, shrubs, conifers and broad-leaved evergreens. The scale is particularly injurious to cultivated trees, causing loss of vigour, branch or crown dieback, and eventual tree death. Trees weakened by walnut scale may become more susceptible to secondary invaders.

Host Plants

Walnut.

General Appearance

The pest is active from May to October. Walnut scale is an armored scale insect that attacks a wide variety of woody plants. The colour of walnut scale varies from white to grey to brown and is often overlooked because it blends well with the colour of walnut bark. This scale insect is often found in groups. Deformity results when male crawlers settle under the margin of female cover and start forming elongate dome shaped covers. The cover of the adult is circular and body is yellowish. Walnut scale overwinters as second instar females and males, resumes development in spring and mate. After mating female lays eggs in mid May and eggs hatch in 2 to 3 days. Soon after hatching crawlers leave the shelter and settle on the branch or twig, and start feeding. Initially the scale cover is white (White cap stage) but changes to brown after about a week (Black cap stage). Walnut scale has two generations a year. The first generation completes in mid July and females start laying eggs in mid August.

Nature of Damage

The scale feeds on the plant sap on stem and branches and cause-stunted growth. The damage of scale cause loss of vigour, branch or crown dieback and eventually tree death. Armored scales suck plant juices from the inner bark by inserting their mouthparts into twigs and branches. Trees look water stressed when encrusted with scale insects. Extremely heavy populations can cause the bark to crack; however, walnut scale rarely causes economic damage.

Management

- ✰ Scale infestation can be monitored by visual inspection of the plant. Black electrical tape can be placed around twigs and branches near a scale infestation in order to track the emergence of crawlers. Crawlers will appear on the tape as yellowish and should be observed with hand lens.
- ✰ Small predatory beetle, *Cybocephalus californicus* (Horn) and two parasitic wasps, *Aphytis proclia* (Walker) and *Encarsia perniciosus* (Tower) are prevalent on walnut orchards and help in controlling the pest.
- ✰ HMO @ 2 per cent (2 litres in 100 litres of water) can suppress low to moderate populations during spring. Dimethoate 30 EC @ 100ml or Chlorpyrifos 20 EC @ 100 ml or Quinalphos 25 EC @ 100 ml per 100 litres of water is effective at the time of egg hatching.

9. Walnut Blister Mite

Walnut blister mite, *Eriophyes erinea* Nalepa (Eriophyidae: Acari) also called walnut gall mite is the occasional pest on walnuts. As an arachnid, this pest is more closely related to spiders and ticks than insects. Feeding on the plant's cellular contents irritates the plant tissue, causing the blisters. Mites reproduce rapidly.

Adults only live for about a month and several generations occur each year. Hot, dusty conditions encourage rapid population increase.

Host Plants

Walnut.

General Appearance

The pest is active from April to October. In late summer overwintering females are produced which will not lay eggs until the following year. The adult female mite overwinters in twigs and buds on the dormant walnut tree. Adult mites are very small and cannot be seen without a hand lens. They have a white, slender, striated body with a few long hairs. Immature forms resemble adults but are smaller. Eggs are spherical and pearly white. Eriophyid mites reproduce rapidly. Fertilization occurs when females come in contact with sperm sacs left on the host by males and lays as many as 80 eggs. Most eriophyid mites have a simple life cycle in which they develop through three growth stages: egg, first and second instar nymphs, and adult. Some species have a more complicated life cycle. They alternate between a generation of only overwintering females called deutogynes, and a male female generation, where the females are called protogynes. Alternating generation is more common in eriophyids that feed on deciduous, woody plants, and appears to be an adaptation based on the seasonal changes of the hosts. Adults live for about one month. There may be two to three generations during the summer with both male and female adults being present.

Nature of Damage

Blister mite feeds on the lower surface of leaflets, causing characteristic blister like swellings on the upper surface of leaflets. Yellow to orange felty masses occur in depressions on the underside of blistered leaves. Later in the season, these areas turn brown. The blisters can be large and unsightly, but damage caused by the walnut blister mite is primarily aesthetic. The leaf size remains small in the nursery plants, thus causing stunted growth.

Figure 14: Damage of Blister Mite on Walnut Leaves.

Management

- ☆ Control of this pest is generally not warranted, as damage is mostly aesthetic. Provide the walnut tree with excellent cultural care, as healthy, vigorous walnut trees are better able to withstand mite activity. Avoid the use of broad spectrum, persistent pesticides on or near the walnut trees, as these chemicals have a serious impact on naturally occurring mite predators. Occasionally spray the walnut tree foliage with a strong spray of water. This may knock some of the mites off of the tree while also addressing the dusty conditions which mites prefer. If the mite activity is isolated on just a few branches, pruning off and disposing of infested portions may offer some control.
- ☆ The mite is difficult to manage once it move into the blister forms which it creates by feeding on the leaf tissue in the early spring. By petal fall, the mites lay eggs and remain protected from predators within the leaf blisters. Predatory phytoseiid mites can usually be found feeding on these mites. Most galls cause aesthetic injury and do not kill their host so biological control by these phytoseiid mites prove successful in controlling the pest. The predatory mite *Amblyseius finlandicus* (Oudemans) and *Typhlodromus occidentalis* (Nesbitt) and feed on the exposed mites but do not provide economic control.
- ☆ HMO @ 2 per cent (2 litres/100 litres of water) are effective for overwintering stages. Apply acaricides just at bud burst stage in April like Fenazaquin 10 EC @ 40 ml or Dicofol 18.5 EC @ 108 ml/100 litres of water. Repeat the application of acaricides during mid June if more than 2 blisters are noticed per leaf.

The other insect pests that affect Walnut are: San Jose scale, *Quadraspidiotus perniciosus* (Comstock), Chaffer beetle, *Adoretus Simplex* (Sharp), Codling moth, *Cydia pomonella* (Linnaeus), Stem borer, *Aeolesthes sarta* (Solsky), European red mite, *Panonychus ulmi* (Koch), Hairy caterpillar, *Lymantria obfuscata* (Walker), Two spotted spider mite, *Tetranychus urticae* (Koch) and Shot hole borer, *Scolytus nitidus* (Muller), are discussed under the insect pests of Apple, Plum and Peach.

Chapter 9

Insect Pests of Chestnut and their Management

Chestnut *Castanea sativa* Mill native to Europe and Asia Minor belong to the family Fagaceae. The four main species of chestnut are commonly known as European (*Castanea sativa*), Chinese (*Castanea mollissima*), Japanese (*Castanea cranata*), and American chestnut (*Castanea dentata*). Chestnuts should neither be confused with horse chestnuts of genus *Aesculus,* which are not related to *Castanea* and are named for producing nuts of similar appearance, but are mildly poisonous to humans, nor should they be confused with water chestnut (family Cyperaceae), which are also unrelated to *Castanea* and are tubers of similar taste from an aquatic herbaceous plant. Other trees commonly mistaken for chestnut trees are the chestnut oak (*Quercus prinus*) and the American beech (*Fagus grandifolia*), both of which are also in Fagaceae. Lifespan of chestnut tree depends on the species. It can survive from 200 to 800 years in the wild. Chestnuts have relatively few pests of minor importance. Most commonly managed insect pests are Potato leafhopper, Japanese beetle, Rose chafer and European red mites.

1. Potato Leafhopper

Like many plants, chestnuts are sensitive to the saliva of potato leafhopper, *Empoasca fabae* Harris (Cicadellidae: Hemiptera) that is injected by the insect while feeding. Damage to leaf tissue can cause reduced photosynthesis, which can impact production, quality, and damage the tree. The potato leafhopper feeds on a variety of plant species and has been reported to feed on nearly 200 kinds of plants.

Host Plants

Chestnut, Walnut, Cherry, Loquat, Cherry and Crab apple.

General Appearance

The pest overwinters as eggs or adults. Eggs are deposited in the midrib or larger veins of the leaves, or in the petioles or stems. It is very difficult to detect the eggs. The nymphs hatch in about 7 to 10 days. The nymphs go through 5 instars. The nymphal instars require about 2 weeks to develop into adults. Adults live for about one month, but have been recorded living as long as 120 days. Females mate within 2 days after their final molt and begin laying eggs about 6 days afterwards. The multiple generations keep damaging the new tip growth of the tree. Potato leafhoppers usually are first found in mid May. An entire life cycle can be completed in about 4 weeks, and as many as six generations may occur each year.

Nature of Damage

Most injury occurs on new tissue and terminals with feeding near the edges of the leaves using piercing and sucking mouthparts. Symptoms of feeding appear as whitish dots arranged in triangular shapes near the edges. Heavily damaged leaves show necrotic and chlorotic edges and eventually abscise from the tree. Severely infested shoots produce small, bunched leaves with reduced photosynthetic capacity, dwarfing of plants and stunted growth is also seen. This damage is typically referred to as hopper burn.

Management

- ✰ Scouting should be performed weekly as soon as nymphs appear to ensure early detection and prevent injury.
- ✰ The easiest way to observe potato leafhopper is by flipping the shoots or leaves over and looking for adults and nymphs on the underside of leaves. Pay special attention to succulent new leaves on the terminals of branches.
- ✰ Spray Methyl-*O*-demeton 25 EC @ 80 ml or Dimethoate 30 EC @ 100 ml/100 litres of water when nymph infestation is noticed.

2. Japanese Beetle

Japanese beetle *Popillia japonica* Newman (Scarabaeidae: Coleoptera) feed on more than 300 plant species, including chestnuts. Japanese beetle adults are considered a generalist pest and affect many crops found on or near grassy areas, particularly irrigated turf. The pest originates from North Eastern Asia.

Host Plants

Chestnut, Apple, Plum, Peach, Walnut *etc.*

General Appearance

The pest first appears in late spring or summer and continues to feed and mate for rest of the season. Egg laying begins soon after the adults emerge from the ground and mate. Females leave plants in the afternoon, burrow 2 to 3 inches into the soil in a suitable area, and lay their eggs. Females are attracted to moist, grassy areas to lay their eggs. Around 40 to 60 eggs are laid during their life. The grubs after hatching grow quickly and by late August are almost full sized. The developing beetles spend the next 10 months in the soil as white grubs, feed on the roots of plants and seedlings. Older grubs are relatively drought resistant and will move deeper into the soil if conditions become very dry. Japanese beetle grubs can withstand high soil moisture, so excessive rainfall does not bother them. Japanese beetles overwinter in the grub stage. When the soil cools in the fall, the grubs begin to move deeper. Most pass the winter below the surface of soil at a depth of 2-5 inches; some may go as deep as 8 to 10 inches. They become inactive when soil temperature falls to about 10° C. When soil temperature climbs above 10°C in the spring, the grubs begin to move up into the root zone. Following a feeding period of 4 to 6 weeks, the grubs pupate in an earthen cell and remain there until emerging as adults.

Nature of Damage

Adults eat leaves but not the veins, producing a characteristic, lacy looking damage. Established trees can generally withstand the defoliation, but young trees are severely affected. Grubs eat the rootlets and weaken the tree due to poor absorption of nutrients.

Management

- ✰ On cool mornings beetles can be easily knocked off the plant into a bucket of soapy water mixed with pesticides, which are recommended for Japanese beetles. Although traps are available, they are usually counterproductive because they attract more beetles than they catch.
- ✰ Parasitic nematodes can be applied to soil to reduce the number of overwintering grubs.
- ✰ Keep plants healthy by following a recommended irrigation and fertilization schedule.
- ✰ Start monitoring for grubs in the early spring by taking a few soil samples, and scout for adults in the summer by inspecting plants.
- ✰ Soil application of Carbofuran 3 CG @ 32.5 kg/ha or Quinalphos 10 per cent dust @ 20 to 25 kg/ha helps to control the grub population.
- ✰ If adult feeding damage becomes noticeable, apply Chlorpyrifos 20 EC @ 100 ml/100 litres of water.

3. Chestnut Weevil

The most important insect pest of chestnut trees is the lesser chestnut weevil, *Curculio sayi* (Curculionidae: Coleoptera). Large chestnut weevil *Curculio caryatrypes* is also an important pest, but is less prevalent.

Host Plants

Chestnut.

General Appearance

The chestnut weevils emerge from the ground in late May to July when the chestnuts bloom, but do not lay eggs until the fall. Egg laying begins when the nuts are nearly mature. Eggs are usually laid in the downy inner lining of the brown shell covering the nut. Eggs hatch in about 10 days and grub development is completed in 2 to 3 weeks. Soon after the nut falls to the ground, the grubs chew a circular pinhole in the side of the nut to enter the soil. They burrow few inches into the ground and spend the winter inactive underground. Most of the chestnut weevil grubs overwinter the first year as grubs, pupate the following fall, and overwinter the following winter as adults. Some pass two winters in the grub stage and a third winter as adults before emerging from the ground in search of nuts. The life cycle is completed in 2 to 3 years.

Nature of Damage

The grub burrows into the nuts and eat them from the inside out, destroying them completely. Tunnels in the flesh of the nuts are the main indicator, though the whitish grub may be observed, as well.

Management

- ✰ Picking up the nuts as soon as they fall prevents the grub from migrating into the soil and continuing their life cycle.
- ✰ Drench the soil with Chlorpyrifos 20 EC @ 300 ml/100 litres of water or apply either Quinalphos 10 per cent dust @ 20 to 25 kg/ha or Carbofuran 3 CG @ 32.5 kg/ha to kill the overwintering stages of the pest.
- ✰ Spray in the fall when beetles start to lay eggs with a contact insecticide like Dimethoate 30 EC @ 100 ml/100 litres of water.

The other insect pests that affect Chestnut are: European red mite, *Panonychus ulmi* (Koch), Two spotted mite, *Tetranychus urticae* (Koch), Chaffer beetle, *Adoretus simplex* (Sharp) and Shot hole borer, *Scolytus nitidus* (Muller) are discussed under the insect pests of Apple.

Chapter 10

Insect Pests of Mulberry and their Management

Mulberry is the name given to several species of deciduous shrubs or trees in the genus *Morus* (family Moraceae) that are grown for their edible fruits. The genus includes white mulberry (*Morus alba* L.) and red mulberry (*Morus rubra* L.) Mulberries are small to medium sized trees with light green leaves, which vary in shape depending on variety. The trees produce small green yellow flowers.The fruit can be white, pink purple or black in colour and contains numerous brown seeds. Mulberry can reach to a height of 15 to 20 m but in its cultivated form, is often pruned to a low growing bush to facilitate the harvesting of fruits. Mulberries are quite short lived, with an economic lifespan of around 15 years. It is also used as silkworm feed and timber in many parts of the world. It originated from South East Asia. Over 300 insect pests belonging to various orders are known to attack mulberry and cause damage. They attack mulberry during different seasons, causing severe leaf yield loss and leaf quality. Since, majority of insect pests destroy tender leaves, young age silkworm rearing will be affected which in turn affect the later stage also. The insect pests that affect mulberry are:

1. Mealy Bug

Mealy bug, *Maconellicoccus hirsutus* Green (Pseudococcidae: Hemiptera) is regarded as a serious polyphagous pest feeding voraciously on mulberry leaves. These insects have a greyish to pink body and are covered in a white waxy material. They tend to form masses where they feed on the plant, producing honeydew and impacting plant growth.

Host Plants

Mulberry, Grape, Plum, Crab apple, Pomegranate, Almond *etc.*

General Appearance

Mealy bugs occur throughout the year, but are severe during summer months. Reproduction is mostly parthenogenetic or sometimes-biparental males are reported to have a pupal stage capable of locomotion. Most adult female mealy bugs are wingless, soft-bodied, greyish insects. They are usually elongate and segmented, and may have wax filaments radiating from the body, especially at the tail. Females can move slowly and are covered with whitish, mealy or cottony wax. This waxy covering is similar to that produced by cottony cushion scale. Each adult female lays 150 to 600 eggs over a period of about one week in egg sack of white wax in clusters on twigs, branches and bark. Eggs hatch in 3 to 9 days. Movement occurs at the first instar stage. Crawlers are very small and can survive a day or so without feeding. They disperse through wind. Once the crawler settles at a feeding site development continues; there are three immature instars in the female and four in the male. Crawlers settle in cracks and crevices, usually on new growth, which becomes severely stunted and distorted, in which densely packed colonies develop. The last instar of the male is an inactive stage with wing buds.The nymphal stage lasts for 20 to 30 days. A generation is completed in about five weeks in warm conditions. In countries with a cool winter, the species survives cold conditions as eggs or other stages, both on the host plant and in the soil. There may be as many as 15 generations a year.

Nature of Damage

Mealy bugs tend to congregate in large numbers, forming white, cottony masses on plants. They feed on stems and leaves of fruit trees and ornamentals and injects into the plant a toxic saliva that results in malformed leaf and shoot growth. High populations slow plant growth and cause premature leaf or fruit drop, curling and twig dieback. Mealy bugs can lower fruit quality by covering it with wax or sticky honeydew upon which black sooty mould grows. The deformity symptoms in mulberry leaves are; leaves become dark green, wrinkled and thickened with shortened inter nodal distance resulting in bunchy top appearance or russeting of leaves.

Management

- ☆ Clip off the infested portion by secateur, and destroy by burning. This will help in reducing the chances of recurrence of pest.
- ☆ Natural predators usually provide adequate mealy bug control unless interrupted by ants, dusty conditions or insecticides. If necessary, a strong blast of water or horticultural mineral oil @ 2 per cent (2 litres in 100 litres of water) during dormant period will control mealy bugs.
- ☆ Ladybird beetle *Cryptolaemus montrouzieri* (Mulsant) feed on mealy bugs, and can be used to control an infestation. Mealy bugs can also be controlled using the fungus *Lecanicillium lecanii* (Zimmerman).
- ☆ Diatomaceous earth can be applied around the stem of the plant. Diatomaceous earth contains small silica particles that are trapped within the joints of ants. They cause irritation and eventual death. Diatomaceous

earth is especially useful for an infestation that has developed a symbiotic relationship with local ants

- ☆ Spray Methyl-*O*-demeton 25 EC @ 80 ml or Dimethoate 30 EC @ 100ml/100 litres of water if infestation is in severe form.

2. Brown Hairy Caterpillar

Brown hairy caterpillar, *Porthesia scintillans* Walker (Lymantridae: Lepidoptera) is a serious pest of mulberry. Young caterpillars are gregarious and they feed on the soft and tender tissues, leaving behind only veins. Skeletonized mulberry leaves and black faeces of the caterpillar on the leaves at the lower portion of the mulberry plant indicate infestation by the pest.

Host Plants

Mulberry, vegetables *etc.*

General Appearance

Adult is yellowish with spots on the edges of forewings. The female lays eggs in groups on lower surface of the leaves. The egg period is from 4 to 10 days. There are six larval instars. The larval periods lasts 13 to 29 days. The caterpillar pupates in a silken cocoon in leaf folds for 9 to 25 days. The caterpillar overwinters during winter season.

Nature of Damage

Gregarious young caterpillars feed upon the chlorophyll layer of leaf exposing the veins. Later instar caterpillars are voracious eaters of mulberry leaves. The affected leaves look dead, dried and easily fall off. Clear branches without leaves can also be noticed after a severe attack.

Management

- ☆ Collection and destruction of egg masses and gregarious young larvae. Deep ploughing and flood irrigation help in the reduction of pest.
- ☆ Spray Dichlorvos 76 EC @ 0.7ml/litre of water. Avoid plucking of leaves for 15 days after spraying.
- ☆ To control early stage larvae, spray Chlorpyrifos 20 EC @ 100 ml or Quinalphos 25 EC @ 100 ml/100 litres of water.

3. Leaf Webber

The leaf webber, *Diaphania pulverulentalis* Hampson (Pyralidae: Lepidoptera) is one of the serious pests which infests mulberry. The incidence of this pest becomes severe during rainy and winter seasons. The loss due to this pest is around 30 per cent. The young larvae feed on the tender unopened leaves and they bind or roll the leaves by secreting silky threads.

Host Plants

Mulberry.

General Appearance

Incidence of the pest starts with the onset of monsoon. It occurs from June to February but reaches peak during September to October. The larva binds mulberry leaf blades by silken thread, stay inside and feed. Its fecal matter can be seen below the infested portion. The female moth lays 150 to 200 eggs at the rate of 1 to 2 eggs per apical shoot of mulberry plant and they hatch into larvae after 4 days. The larvae have 5 instars, which last for 15 days and then pupate into the soil or in dry leaves. Pupal stage lasts for 9 to 10 days. The total life cycle completes within a month.

Nature of Damage

The pest causes damage by folding the leaves and by webbing the tender shoots. Larvae web the leaves together and feed from inside on soft tissues, and skeletonize them. Grown up caterpillars feed voraciously on tender leaves. Apical tips are preferred for feeding, resulting in stunting. Also, apical shoots are destroyed due to egg laying. Quality of leaf and yield is severely affected.

Management

- ✰ Collection and burning of dried leaves from the infested area and deep ploughing followed by flood irrigation to expose the hidden pupae to their natural enemies like birds is the best method to reduce the emergence of moths.
- ✰ Installing light traps in the mulberry gardens attracts and kills the moths. The larvae can also be collected and destroyed mechanically by hand picking.
- ✰ Release of the larval parasitoid, *Bracon bravicornis* (Wesmael) @ 200/acre 15 days after pruning. Release *Trichogramma chilonis* (Ishii) egg parasitoid @ 1 Tricho card/week (for 4 weeks). Do not spray any insecticide after the release of *Trichogramma* parasitoids.
- ✰ Other natural enemies like *Apanteles* spp. (nr. Nepitae Wilkinson) and *Chelonus* spp. (nr. Curvimaculatus) were also found to be parasitic on leaf Webber.
- ✰ Spraying of Chlorpyrifos 20 EC @ 1ml/litre of water two times at the interval of 10 days is recommended in case of severe infestation.

4. Oriental Leaf Worm Moth

Oriental leaf worm moth *Spodoptera litura* Fabricius (Noctuidae: Lepidoptera) is a polyphagous insect also known as tobacco budworm or tobacco cutworm. It occasionally damages the mulberry plants.

Host Plants

Mulberry and other fruits trees.

General Appearance

Females lay eggs in clusters of 200 to 300 underneath the leaves for a period of

6 to 8 days. They are covered with body hair scales. The incubation period is from 4 to 5 days. After emergence caterpillar starts feeding and eat entire leaves, flowers and fruits. The larvae go through six instars. The larval period lasts for 2 to 3 weeks. Caterpillar pupates in the soil in an earthen cocoon. Adult emerges from pupa in about two weeks. Life cycle is completed in 30 to 40 days.

Nature of Damage

The caterpillars attack the shoot of young plants and cut them. The cut portion of the shoot dries up and falls down. They also feed upon the leaves gregariously by scraping the leaves. They are first gregarious and later solitary. Newly sprouted mulberry garden having young plants are found with cut branches with dried leaves.

Management

- ☆ Deep ploughing of mulberry garden exposes the different stages of pest, which can be picked up and killed.
- ☆ After pruning, deep digging around the mulberry plants and application of carbofuran 3 CG @ 32.5 kg/ha or dusting with Quinalphos 10 per cent dust @ 25 kg/ha kills the caterpillars. Leaves can be utilized 45 days after the dusting.
- ☆ Spray Dimethoate 30 EC @ 100 ml/100 litres of water on mulberry plants also kills the caterpillars.

The other pests that affect Mulberry are two-spotted mite *Tetranychus urticae* (Koch), and scale insect *Quadraspidiotus perniciosus* (Comstock) are discussed under the insect pests of Apple.

Chapter 11

Insect Pests of Strawberry and their Management

Strawberry is the name given to several plant species in the genus *Fragaria* a group of flowering plants that include both domesticated and wild strawberries, which are grown for their edible fruit. Strawberry, *Fragaria* spp. L. is an herbaceous perennial plant in the family Rosaceae. Strawberry can grow 8 to 10 inch in height and has an economic life of 2 to 4 years before the plants are replaced. Strawberry may also be referred to as garden strawberry and the plant is grown in most northern temperate regions of the world. Strawberry plants are susceptible to threat from various insect pests. A number of precautionary and protective measures are required in order to achieve maximum production. The following insect pests are found on strawberry plants.

1. Strawberry Fruit Fly

Strawberry fruit fly *Drosophila melanogaster* Meigen (Drosophilidae: Diptera) are small, yellowish flies, 2 to 4 mm in length that are commonly attracted to fermenting fruit of all kinds. Females are slightly larger than males. The fly is found in all temperate regions of the world. The scientific name *Drosophila* means lover of dew. Maggots are about 1/4 inch long and can be found in ripe and damaged fruits in the fields.

Host Plants

Strawberry.

General Appearance

The life cycle is approximately 70 to 80 days during winter, with adults laying about 26 eggs per day or 700 to 800 eggs in a 20 to 30 day life span on a piece of fermenting fruit or other decaying sweet organic material. Populations build up as temperatures become warmer. The flies do not lay eggs at temperatures below 12°C or above 33°C. The adult Drosophila may live for more than 10 weeks. During this time, mating takes place. Females reach the peak of their egg production between the 4th and 7th day after their emergence. During this time, they lay eggs almost continuously at a rate of 50 to 70 eggs per day. The eggs are approximately one-half mm in length, white, oval, and slightly flattened. Eggs are laid half-buried in rotten fruit. Eggs hatch in 22 to 24 hours.The maggot that emerges looks like a tiny worm and is called the first instar maggot. It feeds on the substrate that the eggs were laid in and, after another 25 hours, moult into a larger wormlike form, the second instar maggot. After about 24 hours, moult into the third instar maggot. This is the largest of the larval forms. It feeds, but it also starts to climb upward out of its food, so that it will be in a relatively clean, dark and dry area to undergo pupation. The third instar moult into a pupa after 30 hours. The pupa is stationary, and in its early stage is yellowish-white. As it develops, the pupa becomes progressively darker. The pupal stage lasts for 3 to 4 days, after which the adult fly, or imago, emerges from the pupal case. Adult male flies are sexually active within 48 hours of emerging, females don't have ripe eggs until two days after eclosion, and the cycle begins again. Female fruit flies can store sperm from multiple inseminations for use in future egg production.

Nature of Damage

Fruit flies are attracted to very ripe or damaged fruit in the field where they lay their eggs. They are primarily a problem in strawberries picked for freezing. Because this fruit is allowed to ripen in the field in order to allow easy removal of the strawberry calyx and core during picking, the harvest interval is increased and the fruit becomes more susceptible to infestation. Fruit fly eggs and larvae are primarily a contamination problem.

Management

- ☆ Remove the ripe fruits from the field and bury them. Follow good cultural practices around the field. Practice good sanitation in and around the field. Monitor the flies with sticky traps to help detect the infestation as early as possible. Yellow sticky cards can be used to monitor adult fly populations. Adults and their offspring may also be monitored using fermented fruit traps consisting of a container filled with overripe fruit covered with an inverted funnel.
- ☆ Fly eggs and maggots in the berries cannot be killed using insecticides. Apply treatments to target adult flies. Adult flies are most active in the early morning and late afternoon; this is also the time they will have greater exposure to an insecticide application. The chemicals which are used to control the flies are Malathion 50 EC @ 140 ml or Dichlorvos 76 EC @ 70 ml/100 litres of water.

2. Strawberry Bud Weevil

Strawberry bud weevil *Anthonomus signatus* Say (Curculionidae: Coleoptera) is one of the most destructive strawberry pests. Strawberry is the primary host of bud weevil, but it has been observed damaging raspberry crops as well. The pest has been seen ovipositing in developing buds in strawberries. This dark reddish brown weevil is about 1/10 inch long with a head elongated into a slender, curved snout about half as long as the body. Its back has two large black spots. This insect is known as the strawberry clipper because of its habit of clipping buds. Larvae are tiny creamy white coloured grubs found inside unopened flower buds. The strawberry bud weevil is probably one of the most important direct pests of strawberry. This pest has been shown to cause yield losses from 50 to 100 per cent in some areas.

Host Plants

Strawberry, Raspberry, Blackberry, Cranberry *etc.*

General Appearance

Strawberry bud weevil has one generation per year. The pest overwinters as an adult 3 mm in length, emerging from ground litter commonly in wooded areas and migrates to strawberry fields early in the season when the temperature increases. Ovipositing females puncture unopened buds with their long beaks and deposit a single egg into the bud and then partially clip the blossom below the bud. Eggs hatch in one week. Grubs develop in the severed buds and reach maturity in 3 to 4 weeks. They then pupate in the bud and after 5 to 8 days adults emerge in June, feed on flower pollen and leaves, then enter aestivation in July to August and remain inactive for the rest of the season. The adults have been found mating throughout the oviposition period.

Nature of Damage

Damage results when females sever the strawberry bud from the pedicel following oviposition, causing it to hang by part of the stem, or fall to the ground thus, preventing fruit formation Adult clippers first feed on immature pollen by puncturing nearly mature blossom buds with their snouts. The female then deposits a single egg inside the bud and girdles the bud, preventing it from opening and exposing the developing grub. The adult female then clips the stem so that the bud hangs down or falls to the ground. Grubs feed within the damaged bud for a period of 3 to 4 weeks and a new generation of adults emerges in late June and July. These weevils feed on the pollen of various flowers for a short time, but seek shelter in late summer in preparation for overwintering.

Management

- ✰ Strawberry beds should be located some distance from woods, which provide an overwintering site. Likewise, beds should be free of grass and weeds. Avoiding field site selection near wooded areas prevents high numbers of overwintering adults from entering the field in the spring.

Mulches and full canopy beds encourage adults to overwinter and remain in the field, thus plowing of old beds immediately following harvest causes adult mortality.

- ☆ Topping of plants and removal of foliage and mulch immediately following harvest should be done and then applying a follow up chemical spray to kill overwintering adults.
- ☆ Applications of Malathion 50 EC @ 140 ml/100 litres of water should begin when cut buds are first observed. Applications should continue through the unopened bud period as long as damage persists. As weevils feed during the day, early morning applications have proven to be more effective.

3. Strawberry Leaf Roller

Strawberry leaf roller *Ancylis comptana* Frolich (Tortricidae: Lepidoptera) is a reddish brown tortricid moth introduced from Europe in late 1800s. It is considered a minor or occasional pest of strawberry because of leaf tissue damage. Larvae use silk to spin webs and roll leaves while consuming leaf tissue.

Host Plants

Strawberry, Raspberry, Blackberry.

General Appearance

Strawberry leaf roller has two or three annual generations, the pest overwinter as larvae in folded leaves or leaf litter. Overwintering larvae form pupae from which the first generation of moths emerge and lay eggs. Moths live for about two weeks. Adults are small, tortricid moths that are rust brown in colour with markings of light yellow on the wings. Eggs are laid singly on the undersides of leaves; a female lays 20 to 120 eggs in her lifetime. After an incubation period of 5 to 17 days, first instars feed on leaf tissue and spin a silky covering near the base of the leaf between two prominent veins. The covering is enlarged as larvae grow, and larvae may move and start building a second covering on the same leaf. Eventually, larvae migrate to the top of leaves and begin to fold and roll them together. Mature larvae reach a length of half inch. Pupae are formed within the leaf roll. The average duration of the entire larval period is 24 days and the pupal period is 6 to 18 days. Overwintering larvae undergo diapause induced by shorter day length and lower temperature in early fall. Diapause is completed in about 3 months.

Nature of Damage

Damage results from larvae feeding on foliage and rolling the leaves along mid veins by means of silk webbing and thereby reducing the yield. Earlier larvae feed on unrolled leaves as larvae mature they fold the leaves in half or web them together. Once enclosed in the rolled leaf, larvae continue to feed. Some leaf rollers may consume whole leaf tissue. Leaf feeding results in reduced runner formation, interference with ripening fruit, and kill plant.Strawberries are quite tolerant to the leaf feeding species and can support high population levels without economic loss.

Management

- ☆ Remove accumulated dead plant material where larvae may overwinter, and if feasible, remove leaves that show signs of damage.
- ☆ *Bacillus thuriengenesis* Balsamo and Spinosad sprays are both effective in treating young larvae. These are organic insecticides that have minimal impact on the environment. A decision to apply insecticides should be based on scouting for damage or results of temperature models used to predict occurrence of egg hatch in the spring. First instars do not roll leaves, so are difficult to observe.
- ☆ A parasites *Macrocentrus ancilovorus* Roh usually kill a high percentage of leaf roller larvae.
- ☆ Spray Dimethoate 30 EC @ 100ml/100 litres of water when leaf rolling is observed in huge number.

The other pests that affect Strawberry are Strawberry root weevil *Nemocestes incomptus* (Horn), Strawberry crown borer *Tyloderma fragariae* (Riley), Chaffer beetle, *Adoretus simplex* (Sharp), Cutworm *Agrotis ipsilon* (Hufnagel), and two spotted spider mite *Tetranychus urticae* (Koch) that are of minor importance.

Chapter 12

Insect Pests of Raspberry, Blackberry and their Management

Raspberry, *Rubus occidentalis* Linnaeus also called black raspberry and Blackberry, *Rubus fruticosus* Linnaeus or European blackberry are perennial shrubs in the family Rosacea that are grown for its aggregate black and red fruit of the same name. They often have thorns, but some varieties are thorn less. The leaves alternate along the stem with each group of leaves consisting of 3 to 5 leaflets. The leaves are prickly, bright green, and are toothed along the edges. The life span of shrubs is variable, but they usually live for less than ten years reaching heights of up to 3 m (10 feet).

1. Raspberry Beetle

The raspberry beetle *Byturus tomentosus* De Geer (Byturidae: Coleoptera) also called raspberry fruit worm is a serious pest of raspberries and blackberries. Up to 50 per cent of raspberry harvest may be lost due to this beetle. The beetle causes damage in its grub stage, which feeds at the stalk end of the fruits. It is mainly a problem on summer fruiting raspberries. Early fruits on autumn raspberries may be damaged, but those ripening after late August are less likely to be affected

Host Plants

Raspberry, blackberry, Strawberry, *etc.*

General Appearance

The adult beetle can cause extensive damage if they are in large numbers and attack raspberries or hybrid berries before blossom time. The adult raspberry beetle is between 0.15 and 0.2 inch in length, reddish brown in colour with short hairs covering its whole body. Overwintering in the soil, the adult emerges in the spring between mid April to mid May. The adults feed on leaves, but in areas of high populations they also feed on upper florescence. Adults are then attracted to flower buds and blooms where mating and, subsequently, oviposition occurs. One female may lay one hundred or more eggs, usually one per flower in the month of May to mid July. Eggs hatch out after about ten days. Grubs emerge from the eggs on the flower or immature fruit and begin feeding on the receptacle. The grubs remain in the receptacle until the fruit is harvested or falls to the ground at fruit maturity in late summer. Pupae are present in the soil in late summer to early autumn. Adult beetles emerge from the pupae in late autumn and remain in the soil until the spring. One life cycle is completed per year.

Nature of Damage

The larvae initially feed at the base of the berries but later feed in the inner core or plug. Buds can be destroyed completely, open blossoms are also injured. Ultimately, there will be many small malformed fruits and heavy crop losses. The damage caused by the grubs is even more important. Attacked drupelets turn brown and hard, particularly at the stalk end of the berry. The presence of the grub inside the fruit renders the fruit inedible.

Management

- ✰ The use of a beetle trap can greatly reduce this problem, by attracting and trapping the adult beetles as they emerge from the soil in the spring.
- ✰ Ideally best control will be achieved by spraying raspberries with Dichlorvos 76 EC @ 70 ml/100 litres of water when the first pink fruits are seen and a second application two weeks later, Apply Carbofuran 3 G @ 32.5 kg/ha or spray Phosalone 35 EC @ 140 ml/100 litres of water.

2. Aphids

Large European raspberry aphid *Amphorophora agathonica* Hottes, Large blackberry aphid *Amphorophora rubi* Kaltenbach, Small European raspberry aphid *Amphorophora idaei* Borner, Blackberry aphid *Macrosiphum funestum* Macchiati and Small raspberry aphids *Aphis rubicola* Qestlund (Aphididae: Hemiptera) are destructive soft bodied insects that feed on raspberry and blackberry leaves. The pest is native to North America. Large raspberry aphid *Amphorophora agathonica* feed on the underside of leaves, close to the tip of the raspberry cane, and spread mosaic virus complex to raspberries. Small raspberry aphid *Aphis rubicola* carries the raspberry leaf curl virus. Aphids also produce a substance called honeydew that attracts ants and causes the growth of sooty mold, both of which affect the quality of the berries the plant produces.

Host Plants

Raspberry, blackberry, Strawberry *etc.*

General Appearance

Raspberry aphids overwinter as eggs on brambles. Eggs hatch in the spring. Development is rapid during the summer with many generations. Most aphids reproduce asexually most or all of the year with adult females giving birth to live offspring often as many as 12 per day without mating. Young aphids are called nymphs. They molt, shed their skin four times before becoming adults. There is no pupal stage. Some species produce sexual forms that mate and produce eggs in fall or winter, providing a more hardy stage to survive harsh weather and the absence of foliage on deciduous plants. In some cases, aphids lay these eggs on an alternative host, usually a perennial plant, for winter survival. When the weather is warm, many species of aphids can develop from newborn nymph to reproducing adult in seven to eight days. Because each adult aphid can produce up to 80 offspring in a week, aphid populations can increase with great speed.

Nature of Damage

Aphids live together in colonies on raspberry plants, specifically on the growing tips of young raspberry canes and under raspberry leaves. Aphids feed on raspberry leaves and may cause the leaf petioles to twist or curl. The two species of aphids that feed on raspberry bushes are also carriers of raspberry bush viruses. Large raspberry aphids (*Amphorophora agathonica*) feed on the underside of leaves, close to the tip of the raspberry cane, and spread mosaic virus complex to raspberries. Small raspberry aphids (*Aphis rubicola*) carry the raspberry leaf curl virus, of which black raspberry bushes in particular are highly susceptible.

Management

- ☆ Overwintering aphid eggs are frequently laid on the undersides of leaves. Where practical, burying, removing, or destroying fallen leaves during the winter may decrease the number of colonizing aphids in the early spring. Elimination of wild brambles will reduce migration into cultivated plantings.
- ☆ Aphids have many natural enemies, including ladybirds, hoverfly larvae, lacewing larvae and parasitic wasps.
- ☆ Chemicals like Methyl-*O*-demeton 25 EC 80 ml or Dimethoate 30 EC 100 ml/100 litres of water can be sprayed when there is severe infestation of the pest.

3. Raspberry Sawfly

The raspberry sawfly *Monophadnoides geniculatus* Hartig (Tenthredinidae: Hymenoptera) is an uncommon pest of raspberry. The fly is small, black, wasp like insect that appears in early summer and damages raspberry leaves during its larval stage. The female lays eggs on the leaves. The larvae are small, green in colour and look like little caterpillars. They feed on the leaves, while avoiding the larger veins. The result is a leaf that is peppered with small holes, creating a distinct netted look, a type of feeding referred to as skeletonizing.

Host Plants

Raspberry, Blackberry.

General Appearance

Raspberry sawfly adults are tiny stout black wasps measuring 6 mm in length and emerge in spring when there are blossoms in the field. They lay eggs in raspberry leaves. Eggs are inserted in leaf tissue. Larvae are bristle and pale green and eventually reach 12 mm in length. Larvae are very similar in colour to the leaf and generally feed on the underside of leaf for 2 to 3 weeks during summer. They are often present in groups. Larvae feed mostly between the veins, which causes long holes and shredding. In severe infestation, leaves are skeletonized, however, in most cases vigorous raspberry plantings can easily tolerate obvious feeding from sawfly larvae. After feeding they drop to the ground where they overwinter in cocoons. The pest pupates in the spring. Adult sawflies are present in June and sporadically through the summer. Adults are small, thick bodied wasps. There is one generation per year but sometimes there may be a partial second generation.

Nature of Damage

The pest feed on underside of leaves, flower buds, young fruit and tender bark of growing shoots. The initial damage appears as small holes in leaves and completely devours the leaves leaving elongated holes between the larger veins or sometimes leaving only framework of leaves. The feeding by the pest results in weak plants.

Management

- ☆ Cultivate around trees and shrubs in the early spring and again in the fall to help reduce the overwintering population.
- ☆ During July, watch for holes in the leaves. The pest is often controlled when spraying is done for other insects.
- ☆ Spray Dimethoate 30 EC @ 100 ml or Phosalone 35 EC @ 140 ml/100 litres of water when there is infestation of the pest.

4. Spotted Wing *Drosophila*

Spotted wing drosophila *Drosophila suzukii* Matsumura (Drosophilidae: Diptera) is a polyphagous pest and have been found infesting many fruits including berries. The fly originated from Asia. *Drosophila suzukii* adults are small yellowish brown flies with red eyes. The adults have a pale brown or yellowish brown thorax with black bands on the abdomen. Males have a distinguishing dark spot along the front edge of each wing. Spotless males are also possible, but are rarely observed in the field.

Host Plants

Raspberry, Blackberry, Strawberry, Cherry, blueberry, Plums, Grape, Nectarines, and Fig.

General Appearance

Flies emerge in spring, but some adults may be active during warm winter days. Eggs are laid in ripening fruits, females use their ovipositors to cut the surface of fruit and lay approximately 1 to 3 eggs per fruit and 7 to 15 eggs per day. The egg laying lasts 10 to 15 days. The eggs are translucent, milky white, and glossy. Each female can lay 195 eggs during her lifetime.The eggs develop and hatch within the fruit in which they are laid. Eggs hatch in 1 to 3 days, maggot mature in 3 to 13 days and most of them pupate in the fruit, but some drop and creep into the soil. First instar maggots are approximately 0.07 mm in length. Maggot development occurs inside the fruit and develops through three instars before pupation. Mature maggot may grow up to 6 mm in length. After 5 to 7 days of maggot feeding inside, the skin collapses and fruit leak the juice and begin to rot. Maggots then exit the fruit to pupate and take 3 to 15 days for adult to emerge. Pupation can occur either inside or on the exterior of fruit. The life cycle from egg hatching to adult emergence ranges from 20 to 30 days. The lifespan of adults is 20 to 56 days, but some overwintering adults live for more than 200 days. Around thirteen generations are completed per year.

Nature of Damage

The fly burrows through the berries, making the fruit soft and unappealing. During egg laying the female may introduce fungi that cause the fruit to rot, and infested fruit often develop a fermented or a sour smell. If berries are stored at room temperature, maggots can hatch after picking, causing raspberries that looked normal during picking to deteriorate a few hours later. Populations are highest between the middle of July and the first frost, so both summer bearing raspberries and fall bearing raspberries can be destroyed. During severe infestations, nearly all the fruits will have maggots.

Management

- Eliminating any fruit that has fallen on the ground and any infested fruit remaining on plants in the garden can reduce populations of flies that might infest next year's crop or later ripening varieties. Early harvest of fruit can be important in reducing exposure of fruit to the pest. Begin harvest as early as you can and continue to remove fruit as soon as they ripen.
- Prune plants to maintain an open canopy. This may make plantings less attractive to fly and will improve spray coverage. Allow the ground and mulch surface to dry before irrigating.
- Traps are typically used to detect adult flies and determine whether control measures are needed. Spray Phosalone 35 EC @ 140 ml or Ethion 50 EC @ 100 ml/100 litres of water.

The other pests that attack Raspberry and Blackberry are Crown borer *Pennisetia marginata* Harris, Rose Chafer *Macrodactylus subspinosus* Fabricius, Japanese beetle *Popillia japonica* Newman, Two spotted mite *Tetranychus urticae* Koch, and Cane borer *Oberea perspicillata* Haldeman that are of minor importance.

Chapter 13

Insect Pests of Grape and their Management

Grape *Vitis vinifera* Linnaeus is one of the important fruit crop of world used for table purpose, raisins and wine making with good medicinal value due to the presence of large amount of antioxidants. Insect pests are the important production constraints in grape cultivation. In grape, 85 species of insect pests have been reported in India. Among them, the pests listed below causes serious damage to the vineyards.

1. Grapevine Stem Girdler

Grapevine stem girdler *Sthenias grisator* Fabricius (Cerambycidae: Coleoptera) is a stout beetle, which girdles the vine resulting in drying up of regions above the cut in which the grub can tunnel easily. Discoloration of leaves leads to reduction in photosynthesis thereby affecting the vigour of the plants. Severe infestation of the pest results in delay in maturing and ripening of bunches and reduction in sugar content thereby affecting the quality of grapes.

Host Plants

Grape, apple, citrus, almond and mango.

General Appearance

Adults become active at night and lay oval 1 mm wide and 4 mm long eggs in clusters of 2 to 4, underneath loose bark of girdled branches and hatch in 8 days. Grubs are dark brown and immediately after hatching tunnel into the wood. Grub pupates within the tunnel. Adults are medium sized, and are active at night. Single generation takes more than one year to complete. The grub bores into stem and branches and causes drying and withering of affected branches. Initially reddish sap oozes from wounds; chewed particles of wood are seen on the ground just below the site of damage. The adult beetles start emerging from the vines during July to September by making a round hole on the vine. Female beetles make conspicuous slits on the bark of the trunk and arms of the vine. Presence of sawdust like substance under the vine indicates the damage done by the grub. Damaged vines get weakened and growth gets affected. The maturity of berries is also delayed which ultimately affects the grape production in terms of both yield and quality.

Nature of Damage

Damage is done by female beetle which girdles the branches around the main stem 15 cm above the ground level at night and inserts whitish spindle shaped eggs singly into the tissue in a slanting manner. During the day the beetle hide on the lower side of leaves under the forking of branches. The girdling by the female causes the terminal growth of the new shoots to bend over above the upper girdle and drop to the ground. Later the whole infested shoot dies back to the lower girdle and falls from the vine. Vines attacked by the grape girdler have a ragged appearance suggesting serious injury to the plant. Girdling of the terminal growth has little or no effect on the crop unless fruit producing nodes are close to attacked shoot tips.

Management

- ✰ Collection and destruction of damaged parts should be done. Remove loose bark at the time of pruning to prevent egg laying by female beetle. Cutting the vine one inch below the girdling point and burning them is common method to reduce the infestation.
- ✰ A piece of cloth soaked in an insecticide solution like Chlorpyrifos 20 EC and then wrap around the stem suffocates the grub. Apply cotton impregnated with Dichlorvos 76 EC or one Celphos tablet (3g Aluminum phosphide)/bore hole. Apply Carbofuran 3G (5g/bore hole) and plug with mud. Swab the trunk with Carbaryl 50 WP @ 2g/litre of water.
- ✰ Injection of Formalin 4 per cent solution into bore holes after removing the webs and subsequently sealing of the holes with clay gives satisfactory control of the pest.

2. Grapevine Flea Beetle

Grapevine flea beetle *Scelodonta strigicollis* Olivier (Chrysomelidae: Coleoptera) also known as the steely beetle is found in all grape growing areas of World. This insect primarily attacks buds of wild and cultivated grapevines. The beetle feed on the sprouting buds and eat them completely resulting in poor vine development and improper vine spreading. The adults are very destructive when the vines put forth new flush after pruning. The body of the beetle is somewhat oval in shape, is a metallic shining blue, and measures from 4 to 5 mm in length. The antennae are thread like and about half as long as the body. The hind legs are enlarged, enabling the adult to jump quickly when disturbed as it derives its name from its ability to jump.

Host Plants

Grapevine.

General Appearance

The grape flea beetle overwinters as an adult. Female lays 200 to 550 eggs, beneath bark or in soil in groups of 20 to 40. Egg period is from 4 to 8 days. Eggs are cylindrical in shape, rounded on the ends, yellow in colour and average about 1 mm in length and 0.4 mm in width. Some eggs are placed on the hardened scales surrounding the buds, but most are laid under the loose bark and near the buds. Some eggs are laid on the upper side of the leaves as foliage develops. Newly hatched grubs are dark brown in colour. When the grubs reach maturity they are light brown and 7 to 9 mm in length. The grub pupates in soil and pupal period is from 7 to 11 days. The grape flea beetle passes through this stage of development in a cell prepared in the soil by the grub at a depth of 12 to 65 mm. The pupae are bright yellow, 4 to 6 mm long. The beetles start their activity mainly from May onwards, though they are seen scraping the sprouting buds in early March also. Life cycle is completed in 50-55 days and 3 to 4 overlapping generations are completed in a year.

Nature of Damage

The adult beetles scrap the sprouting buds after each pruning. Damaged buds fail to sprout. The beetles also feed on tender shoots and leaves, which wither and drop down. Adult beetles feeding on primary buds cause damage, which prevents them from developing into shoots, thus resulting in a decreased yield. The greatest economic loss occurs when beetles feed on buds from the time of bud swell until the first leaf separates from the shoot tip stages. Once shoot growth increases damage caused by the grape flea beetle normally does not affect yield. Although primary damage is caused by adult flea beetle to the developing buds, larval damage can also occur on the foliage and is typically limited to several leaves and vines.

Management

- ✰ Remove the loose bark at the time of pruning and rub the stems with jute to remove the egg masses. Leaf litter and debris in and around grapevines should be removed. Monitor grapevines in the early spring for grape flea beetle activity, although grape flea beetles are active later in the summer. Woodlots and wasteland areas near cultivated vineyards are a possible source of flea beetles and should be cleaned up. This will help to reduce sites for beetles to overwinter. Cultivating between rows may contribute to control of the flea beetle pupae by exposing the delicate pupae to desiccation and death.
- ✰ Shake vines to dislodge adult beetles. Collect into trays containing kerosenated water and destroy them.
- ✰ In particularly heavy infestations, if more than 4 per cent of grape buds are infested, spraying with Carbaryl 50 WP @ 2 g/litre or Imidacloprid 17.8 SL @ 0.3 ml/litre of water are recommended to control this pest. Proper

timing of the insecticide is imperative to provide good control of grape flea beetles.

3. Grapevine Thrips

Grapevine thrip *Rhipiphorothrips cruentatus* Hood (Thripidae: Thysanoptera) is the most destructive pest of Grape vine. The two other species that attack grape vine are *Scirtothrips dorsalis* Hood and *Thrips hawaiiensis* Morgan. Nymphs and adults pose an increasing threat to grape cultivation by causing scab formation and resulting in heavy loss in the vineyards.

Host Plants

Grape vine, Jamun, Guava, Mango, Cashew nut *etc.*

General Appearance

Adults are very small, elongated and fast moving measuring 2 mm in length that appear in March and lay eggs on the underside of leaves by making small slits in the plant tissue. The average numbers of eggs laid by a female are 50 to 55. Hatching takes place in 8 to10 days and nymphs emerge which feed on the underside of leaves by rasping the surface and sucking the oozing cell sap. Life cycle of thrip comprises of four immature stages namely first and second instar nymph, pre pupa and pupa. Nymphs move down to the soil and pupate in the top 10 to 12 centimeters. Nymphs are similar to adults but are without wings. The adults feed like nymphs. Females reproduce with or without fertilization. The fertilized eggs hatch into female and unfertilized into male. The thrips hibernate as pupae in soil from December to March. Total life cycle is completed in about 15 days. Adult thrip live for about 10 days.

Nature of Damage

Nymphs and adults rasp the lower surface of the leaf with their stylets and sucking the oozing cell sap. The thrips also attack blossoms and developing berries. Curling of the leaves is observed in case of heavy incidence. The number of minute spots thereby producing a speckled silvery effect marks the injured surface. Such vines either do not bear fruit or the fruit drops prematurely.

Management

- ☆ Field should be kept sanitized, collect and destroy damaged leaves, fruits and flower. Good cultural practices make the crop healthier and reduce the loss caused by thrips.
- ☆ Install 4 to 5 yellow sticky coloured traps per acre to monitor thrips population. Raise ecofriendly crops *viz.* mustard, sunflower, marigold and coriander around the crop to conserve and encourage entomophagous insects.
- ☆ The green lacewing *Chrysoperla carnea* Stephens helps in controlling the thrips.

☆ Spray *Beauveria bassiana* Balsamo @ 5 ml or 5 gm/litre of water under humid climatic conditions. Apply 500 ml of Malathion 50 EC or 1.5 kg of Carbaryl 50 WP in 500 litres of water per 100 vines. Spray Methyl-o-demeton 25 EC, Dimethoate 30 EC, Quinalphos 25 EC, or Ethion 50 EC @ 1ml/litre of water when needed before flowering and after fruit set.

4. Grapevine Mealy Bug

Grapevine mealy bug, *Maconellicoccus hirsutus* Green (Pseudococcidae: Hemiptera) is a serious pest on grapevine varieties having compact fruit bunches like Thompson seedless. Grape Vine mealy bug was first found in California in the mid 1990s, becoming among the most significant vineyard pests. Mealy bugs are soft bodied sucking insects covered in white filamentous wax. Adult females grow to about 5 mm in length and are wingless whereas males are much smaller and winged.

Host Plants

Grape vine.

General Appearance

Female mealy bugs can lay enormous number of eggs, which quickly hatch into crawlers. In early summer, mealy bugs are present mainly along leaf veins. Up to three to four generations may occur each year depending on climatic conditions. Mealy bugs prefer mild, humid conditions and high rates of mortality can occur during hot, dry conditions. Feeding by mealy bugs does not usually cause economic damage. Excretion of sticky honeydew by mealy bugs leads to sooty mould development on leaves and bunches if large populations arise. Sooty mould covering leaves can reduce photosynthesis and mould on grapes can make the fruit unsellable or lead to rotting.

The mature female lays eggs in an egg sack of white wax, usually in clusters on the twigs, branches, and bark of the host plant, and also on the plant's leaves and terminal ends. Female mealy bugs lay 350 to 500 orange coloured eggs in a loose cottony terminal ovisac for a period of 5 to 10 days. Eggs are minute, varying from 0.3 to 0.4 mm in length which quickly hatches into crawlers. They may disperse over the host, especially toward tender growing parts, or be carried away by wind, man, or animals. The nymphal stages appear much like the female in form, but the female nymphs have three instars, while male nymphs have four instars. The last instar of the male is an inactive stage with wing buds within a cocoon of mealy wax. Both female and male adult mealy bugs are about 3 mm long. They are wingless and appear as ovoid shapes covered by a mass of white mealy wax. Males have a pair of wings and two long waxy tails and are capable of flight. Reproduction may occur by means of parthenogenesis in the absence of the male. The mealy bug can complete its entire life cycle in 23 to 30 days. One generation is completed per month, but life cycle extends in winter months.

Nature of Damage

Both nymphs and adults suck sap that results in crinkling and yellowing of leaves and rotting of berries. As it feeds, the mealy bug injects into the plant toxic saliva that results in malformed leaf and shoot growth, stunting, and occasional death. Leaves show a characteristic curling, similar to damage caused by viruses. Heavily infested plants have shortened internodes leading to russeting or a bunchy top appearance. A heavy black sooty mold may develop on an infested plant's leaves and stems as a result of the mealy bugs heavy honeydew secretions. When fruits are infested, they can be entirely covered with the white waxy coating of the mealy bug. Infestation can lead to fruit drop, or fruit may remain on the host in a dried and shriveled condition. If flower blossoms are attacked, the fruit sets poorly.

Management

- ☆ Removal of weeds and alternate host plants harboring the mealy bugs in and around the vineyards throughout the year. Deep ploughing in summer or raking of soil in vineyards helps to destroy its pupal stages and minimizing the incidence. Application of sticky bands or sticky tapes around the stem prevents crawlers from reaching young shoots.
- ☆ Release of Mealy bug ladybird beetle *Cryptolaemus montrouzieri* Mulsant also called mealy bug destroyer @ 5000 adults/ha or 8 to 10 per vine in August to September helps to clear the mealy bug population present on the plants.
- ☆ Swabbing/washing of trunk with 2 ml of Dichlorvos 76 EC + 2 gm of fish oil soap in a litre of water in April to May reduces mealy bugs infestation.
- ☆ Spray HMO @ 2 per cent (2 litres/100 litres of water) during the onset of spring. Apply Carbofuran 3 CG @ 32.5 kg/ha in the soil.

5. White Gall Mite

White gall mite *Colomerus vitis* Pagenstecher (Eriophyidae: Acarina) are tiny less than 0.3mm long sap sucking white or creamy worm like arachnids that cause a variety of abnormal growths on various plants. White gall mite is a widespread pest of grapevine which causes characteristic swellings and rolling of the leaves as well as damage to the buds. Leaf rolling and bud damage have been attributed to different strains of the mite compared to the common *erinea* causing form. Heavy infestations can result in delayed development of the vines and a reduction in grape production. Body length of mite is 0.18 to 0.22 mm and 0.05 to 0.06 mm wide. Piercing and suctorial type of mouthparts are located on the forepart of body called gnathosoma. Chelicerae curved and needle shaped. Propodosoma is the next part of body, with 2 pairs of legs directed forward. Abdomen or opisthosoma ends with a tail plate bearing 2 long setae on each side. Life cycle of mites includes stages of Egg, Protonymph, Deutonymph, and Adult.

Host Plants

Grape vine.

General Appearance

Females hibernate under the outer bud scales and inside cracks at base of one-year-old sprouts. Feeding on juice begins in spring after bud breaking. The mites are usually located on the lower leaf surface. They cause gall and erineum formation 0.2 to 0.86 cm in diameter, which look like hemispherical swellings, being convex on upper surface of leaf and concave on underside. Galls on the lower leaf surface are covered with a felt like layer, which is white at first and brownish later. The felt like layer consists of curled trichomes being produced by leaf epithelium after damage by mite. At least two biological races of this mite are known. Mites of the 1st one-cause galls on leaves, mites of the 2nd race attack buds. Reproduction is by arrhenotoky where spermatophores are deposited by males on the leaves and taken up by the females. The Grape erineum mite gives 5 to 11 generations in a year. Female lays eggs between trichomes after additional feeding. Fertility is about 40 eggs. Egg develops after 10 to 12 days. First galls appear in the beginning of May. Winter diapause begins in early autumn.

Nature of Damage

The mite cause death of overwintering buds in which they undergo hibernation. The whitish areas appear on leaves followed by felt like or yellow patches and gall formation occurs on upper surface. Adult erineum mites overwinter under the scales of buds. Mites move in spring to developing shoots and create galls on young leaves around the fruiting zone. Several generations are produced each year, with new galls developing higher up the shoots. Beginning in late summer, adult mites move back to the buds for winter. Leaves with a few erineum galls appear to function normally and there is little or no economic damage to mature vines until almost all leaves are covered with galls. Even then, damage usually only occurs if vines are also suffering from other stresses.

Figure 15: Grape Leaves Damaged by White Gall Mite.

Management

- ☆ Removal of the infested leaves and branches. Spraying very high volumes of wettable sulphur and oil to runoff at the time of bud stage most effectively treats grape mite. Sulfur has long been used for mite control. Full coverage is necessary for control. Do not apply during periods of high humidity.
- ☆ Biological control using indigenous predatory mites, particularly phytoseiids, has proven successful against gall mite. A natural predator, *Glaendromus occidentalis* Nesbitt feeds on erineum mites. Introduction of this predator has some effect on reducing their numbers; however, the dense hairs of the galls often protect the tiny mites.
- ☆ Apply dormant horticultural mineral oils 7 to 10 days before bud break and again at bud break. Proper timing targets eriophyid mites and preserves beneficial arthropods. During the summer, oils offer mite control or suppression. Proper timing can also target other common plant pests such as aphids.
- ☆ Predatory mites are susceptible to several insecticides, so chemicals that are friendly to predatory mites should be selected to ensure high number of predatory mites in the vineyard. Spray Fenazaquin 10 EC @ 40 ml or Dicofol 18.5 EC @ 108ml/100 litres of water.

6. Chafer Beetle

Chafer beetle, *Macrodactylus subspinosus* Fabricius (Scaraebidae: Coleoptera), is commonly found in many grape growing areas particularly with sandy soil. Rose chafer was first reported as a grape pest as early as 1810, later extending its host range to include a wide assortment of host plants. Grape remains among the most severely injured crops. It is a pest on many different types of flowers, fruits, trees and shrubs. The rose chafer has a yellowish tan coloured body that is about 8 to 13 mm (0.3–0.5 in) in length, with wings that do not completely cover the abdomen.

Host Plants

Grape, Apple, Pear, Peach, Cherry, Raspberry, Strawberry *etc.*

General Appearance

The adult chafers become active from late May. Beetles feed and mate soon after emerging from the soil and it is common to see mating pairs in the newly formed grape vines. Adult chafer beetles deposit eggs singly few centimeters below the soil surface. Mating and egg laying occur continuously for about two weeks with each female depositing 24 to 36 eggs in a separate cavity. The average lifespan of the adult is about three to four weeks. Approximately two weeks after being deposited, eggs hatch into grubs. The grubs feed on the roots of grasses, weeds, grains, and other plants throughout the summer, becoming fully developed by autumn. Grubs move downward in the soil as soil temperatures decline and form an earthen cell in which they overwinter. In the spring, grubs return to the soil surface, feed for a short time, and pupate in May. After two weeks in the pupal stage the adults emerge and crawl to the soil surface to begin their cycle again. There is one generation per year.

Nature of Damage

Adult beetles feed on leaves, buds, blossom and fruit let. Eaten away leaves are perforated. The grubs feed on roots and may cause wilting of the plant. The adult rose chafer attacks the flowers; buds, foliage, and fruit of numerous plants Adults emerge about the time of grape bloom and often cause extensive damage to foliage. Blossom buds are often completely destroyed, resulting in little or no grape production. Feeding activity on various plants may continue for four to six weeks. Damage can be especially heavy in sandy areas. Rose chafers show typical damage on leaves by eating the leaf tissue between the large veins, a type of injury known as skeletonizing.

Management

- ✩ Physically remove rose chafers, especially when small numbers are present. Remove them from plants and put into soapy water to kill them, because chafers are good fliers. Deep ploughing after summer showers will expose pupae and beetle to sun and birds, which results in reduction of pest population.
- ✩ Chemical controls for the rose chafer include Chlorpyrifos 20 EC @ 100 ml, Phosalone 35 EC @ 140ml/100 litres of water. Apply Carbofuran 3 CG @ 32.5 kg/ha in the soil against the grubs.
- ✩ Spraying of Carbaryl 50 WP @ 3g/litre surrounding trees or host plants just after monsoon showers have been found effective against beetles.

7. Grape Vine Stem Borer

Grape vine stem borer *Coelosterna scabrator* Fabricius (Cerambycidae: Coleoptera) is the most serious pest of Grapevine. Earlier this pest was considered to be a problem in unmanaged orchards but in recent years severe incidence of this pest have been recorded in managed orchards as well. Adults are 20 to 28 mm in size. The grub bores in to stem and branches and causes drying and withering of affected branches. Initially reddish sap oozes from wounds; chewed particles of wood are seen on the ground just below the site of damage.

Host Plants

Grape.

General Appearance

The adult beetles start emerging from the vines during July to September by making a round hole on the vine. Female beetles make conspicuous slits on the bark of the trunk and arms of the vine. An adult beetle is about 4 cm long and dull yellow with minute spots. Capsule shaped eggs are laid singly in loose bark of stem in slits and the slits are covered with a hard gummy substance. Eggs are laid on young plants on stems and are deposited under the bark. The number of eggs laid by female range from 20 to 40. Newly hatched flat-headed cream coloured grubs burrow into the trunk or arms and feed inside and make them hollow. Newly emerged grubs are about 1/4 of an inch long. The mature grub is about 2.1 inches long. On hatching the grubs feeds on the soft tissues around the oviposition cavity and then bores into the stem and roots.The grub spends upto 10 months in tunnel they excavate up to May. Thereafter, they pupate inside the tunnel made in the vine. Pupal period is 16 to 18 days. One generation is completed in a year and longevity of beetles is 45 to 60 days. The beetle emerges by making a circular hole through the bark.

Nature of Damage

Both adult and grub damage the vine trees. Adult beetles gnaw the shoots and grub bore inside the trunk and feed on sapwood leading to drying from the injury point whereas, shoots with girth size more than 2 cm can tolerate the injury. Presence of sawdust like substance under the vine indicates the damage done by the grub. Damaged vines get weakened and growth gets affected. The maturity of berries is also delayed which ultimately affects the grape production in terms of both yield and quality. Holes on bark of main stems, excreta and dry powdered material are usually seen near the base of plants.

Management

- ☆ Detect early infestation periodically looking out for drying branches. Clean cultivation and maintenance of health and vigour of the tree should be followed.
- ☆ Hand picking of adults should be done during July to September. Destroy eggs or first instar grubs in the slits during July to September by removing the slit portion with the help of a sharp knife. Removal of loose bark in pre monsoon period, later swabbing the bark with neem oil suspension is recommended.
- ☆ Inject Dichlorvos 76 EC @ 3 ml/litre of water or treat the holes produced by borer with Celphos @ 1 tablet/hole and seal holes with mud.
- ☆ Spray all the surrounding trees with Quinalphos 25 EC or Chlorpyrifos 20 EC @ 100ml/100 litres of water.

The other insect pests that attack Grape are: Light brown apple moth, *Epiphyas postvittana* (Walker), Grapevine hawk moth, *Hippotion celerio* (Linnaeus), Berry plume moth, *Oxyptilus regulus* (Meyrick), Leaf roller, *Sylepta lunalis* (Guenee), Grapevine scale, *Parthenolecanium persicae* (Fabricius), and Two spotted mite, *Tetranychus urticae* (Koch) are of minor importance.

Figure 1: Females with Scale Cover Removed to Show Scale Body. (p. 3)

Figure 2: Apple Twig Infested with Scale Insects. (p. 3)

Figure 3: Wooly Aphid Colony. (p. 5)

Figure 4: Webbing of *Panonychus ulmi*. (p. 10)

Figure 5: Adult of *Panonychus ulmi*. (p. 10)

Figure 6: Apple Aphid. (p. 13)

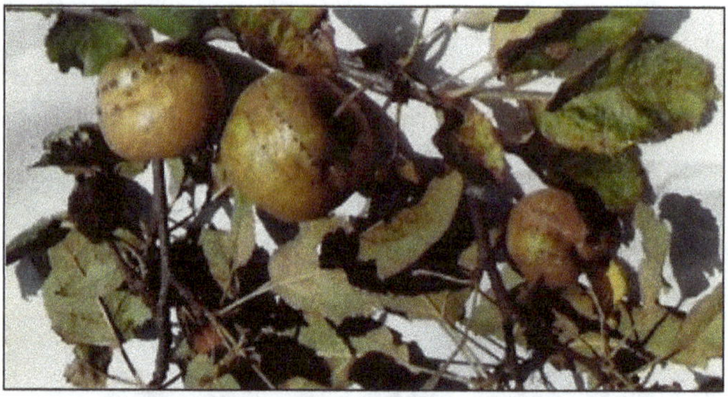

Figure 7: Rust Mite Damage on Apple Leaves and Fruit. (p. 20)

Figure 8: Eggs and Adult of Two Spotted Mite. (p. 22)

Figure 9: Stem Borer Adult. (p. 26)

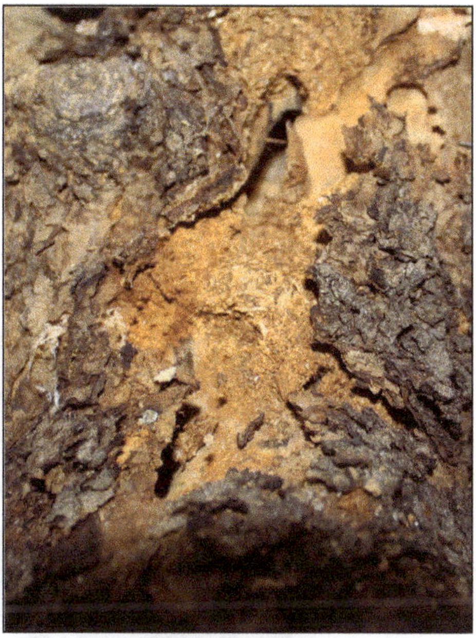

Figure 10: Apple Stem Borer Damage on Apple Branch. (p. 26)

Figure 11: Chafer Beetle Adult. (p. 54)

Figure 12: Eggs of Mealy Bug under Magnification. (p. 64)

Figure 13: Egg Mass of *Drosicha Dalbergiae* in White Cottony Silky Mass. (p. 64)

Figure 14: Damage of Blister Mite on Walnut Leaves. (p. 80)

Figure 15: Grape Leaves Damaged by White Gall Mite. ()p. 111

Chapter 14

Insect Pests of Pomegranate and their Management

Pomegranate *Punica granatum* Linnaeus is one of the commercially important fruit crops of the world as well as India. It is native to Iran. In India pomegranate is cultivated commercially only in the state of Maharashtra. Small-scale plantations are seen in every state of India. Fruit is consumed fresh or in the form of juice, jam, squash and syrup. Among all forms, canned slices and juice are in much demand in India, which consume about 70 per cent of the production. Pomegranate trees are attacked by more than 45 species of insects, which are the major reason behind its low productivity. The major insect pests of pomegranate are as below.

1. Pomegranate Butterfly

Pomegranate butterfly, *Virachola isocrates* Fabricius (Lycaenidae: Lepidoptera) also known as Anar butterfly in India is a polyphagous pest found all over India. It is the most widespread and destructive pest of pomegranate fruit causing forty per cent damage to the fruit.

Host Plants

Pomegranate, Apple, Plum, Citrus, Loquat, Peach, Mulberry and Pear.

General Appearance

The pest remains active from May to September. Butterflies emergence from hibernating pupae from the 3rd week of May. After mating, butterflies lay light green eggs on the flower buds, opened flowers and branches. The eggs are laid singly on flowers and tender fruits. Eggs are shiny white in colour and oval in shape. Incubation period lasts for 7 to 10 days. Soon after hatching, the newly emerged larvae start damaging the flower buds and flowers, then come out for pupation in the 4th week of June. The butterflies from these pupae emerge in the

2nd week of July. The larvae of this generation damage young fruits and bore fruits of pomegranate. In the rainy days of July and August, a complex of fungi grows on the excreta and then extends to the periphery of the hole, finally covering whole of the fruit slowly and sometimes seeds inside get rotten. They develop a characteristic smell and ooze out a dark brown sticky fluid, which attracts the Drosophila species. The 3rd generation butterflies emerge in the 4th week of August and larvae from this generation do not spare even matured fruits from damaging. Larval period lasts for 18 to 47 days. Pre pupa of the last generation start pupating in the fallen leaves and in upper 5 to 7 cm layer of the soil. It goes under hibernation in pupal stage from October to May of the next year. Pupation in earlier generations occurs either inside the damaged fruits or on the stalk holding it. Pupal period lasts for 7 to 34 days. Total life cycle is completed in 1 to 2 months. It has four overlapping generations in a year.

Nature of Damage

The caterpillars bore inside the developing fruits soon after hatching and are usually found feeding on pulp and seeds just below the rind. The single fruit may contain more than one caterpillar. Bacteria and fungi cause the infested fruit to rot. The conspicuous symptoms of damage are offensive smell and excreta of the caterpillars coming out of entry holes, the excreta are found stuck around the holes. Sometimes the holes may also be seen plugged with the anal end of a caterpillar. The affected fruits ultimately fall down and are of no use.

Management

- ☆ Dig or plough around pomegranate tree immediately after harvest or in the early spring to expose the pupae to predatory birds, other natural enemies and sunlight. Collection and destruction of fallen fruits with exit holes reduce the infestation.
- ☆ Spray of Neem Seed Kernel Extract (NSKE 5 per cent) or Neem oil (3 per cent) in the 3rd week of May. Repeat the spray twice in the first week of July at 15 days interval. These neem based insecticides act as ovipositional deterrent.
- ☆ Applications of physical barriers such as covering the thirty to fifty days old fruits with bags of butter paper help reduce infestation.
- ☆ 2-3 periodical releases of egg parasitoid, *Trichogramma chilonis* (Ishii) @ 300000 adults/ha from first week of July onwards can effectively reduce the pest population by parasitizing its eggs. Use light trap @ 1/ha to monitor the activity of adults.
- ☆ Clipping off calyx cup of flowers immediately after pollination will help to reduce the egg load on the fruits and damage level. Growing of resistant varieties can be effective in reducing the fruit borer damage.
- ☆ Spray Dimethoate 30 EC @ 1ml/litre of water in the first fortnight of June or just after initiation of fruit setting. Three sprays of insecticides *viz.*, Deltamethrin 2.8 EC @ 1 ml or Methyl-*O*-demeton 25 EC @ 0.8 ml or Quinalphos 25 EC @ 2.4 ml/litre of water at fortnight interval from 15th

of July. If presence of live holes is noticed one more insecticide already listed may be sprayed in the first fortnight of September.

2. Aphids

Aphid, *Aphis punicae* Passerini (Aphididae: Hemiptera) is a sap-sucking insect of pomegranate and is serious and widespread insect in most pomegranate growing belts. This species is well known for its ability to reduce plant vigour, facilitate the growth of the mould on leaves, and consequently reduce crop quality and yield. Both adult and nymph feed on leaves, inflorescence and can be sufficiently controlled by natural enemies. In addition sooty mould grows on the honeydew on the outer side of fruit, which can be difficult to remove.

Host Plants

Pomegranate.

General Appearance

Aphis punicae is one among the serious pests attacking pomegranate orchards. Young leaves of pomegranate are highly susceptible to aphid attacks. Colonies occur on the leaves from April to September. High humidity favours the multiplication of aphid. Both winged and wingless forms breed parthenogenetically. The nymphal period lasts for 7 to 9 days. Adults live for 2 to 3 weeks and produce 8 to 22 nymphs per day. Entire life cycle takes 22 to 25 days. It has 12 to 14 generations per year.

Nature of Damage

Both adults and nymphs lives in colonies and suck the sap by penetrating the mouth parts into plant leaves causing yellowing and curling of leaves, wilting of terminal shoots, development of sooty mold and pre mature fruit drop. In severe conditions aphids may damage flowers also. The feeding of *Aphis punicae* causes leaf drop, reduction in pomegranate fruit quality and stunted tree growth.

Management

- ✰ Maximize tree health to help trees withstand aphid populations. Aphid numbers increase easily on stressed trees. Collect and destroy the damaged plant parts. Use yellow sticky traps. Cut out aphid infestation if they are concentrated in a few areas of the bush. Burn the debris immediately or double wrap it in a trash bag so the aphids cannot make their way to other plants. A simple jet of water can permanently remove a significant number of aphids as well as their honeydew.
- ✰ Beneficial insects eventually make their way to trees, consuming the majority of natural aphid populations. Green lacewings and ladybird beetles are especially useful aphid munchers. Release first instar larva of *Chrysoperla carnea* (Stephens) @ 15/flowering branch four times at 10 days interval from flower initiation during April. As sprays potentially kill beneficial insects along with aphids, consider even mild sprays as last resort after trying other methods of aphid control. The aphid is attacked

by several endo parasitoids of the family Aphididae the most promising is *Lysiphlebus fabarum* (Marshall).

- ✩ Spraying of HMO @ 2 per cent (2 litres/100 litres of water) during March reduces overwintering stages of aphid. Spray Methyl-*O*-demeton 25 EC @ 80 ml or Dimethoate 30 EC @ 100 ml/100 litres of water for effective management this pest.

3. Mealy Bug

Mealy bug, *Ferrisia virgata,* Cockerel (Coccidae: Homoptera) is an occasional pest of pomegranate. Natural enemies usually keep it below economic injury level. Many mealy bugs like cottony cushion scale *Icerya purchasi* (Maskell), *Planococcus* spp. (Migula), *Maconellicoccus hirsutus* (Green) have also been reported on pomegranate. Mealy bugs are common sap feeding pests that infest a wide range of plants. Mealy bugs weaken plants and excrete a sticky substance (honeydew) on foliage, which allows the growth of sooty mould. Adult females are oval, greyish yellow, 4 to 4.5 mm long, with two longitudinal, sub median, interrupted dark stripes on the dorsum. With this feature the mealy bug is also called striped mealy bug.

Host Plants

Pomegranate, guava, grape, avocado, banana, cassava and cotton.

General Appearance

Mealy bug reproduce sexually with female mating only once. Females lay several hundred yellow to orange eggs in white egg masses that are typically located on the trunk under bark for a period of 20 to 29 days. Each female produces around 70 eggs. The mealy bug has 3 to 5 overlapping generations a year and usually overwinters in the egg stage. Just before or after bud break in March, the eggs hatch in a day and most of the crawlers move out to young shoots, where they nestle between the leaf petioles and the shoots. Female and male nymphs moult 3 and 4 times respectively. In late May to early June, mealy bugs begin returning to the trunks where they mature and immediately begin depositing eggs. The second generation hatches in early June through early July and the crawlers begin migrating to leaves and fruit. Many can be found in the sucker growth at the base of trees. Second generation mealy bugs mature and lay eggs under bark or on fruit beginning in August. If there is no fruit, they return to the trunks and lay eggs that will overwinter. Longevity of the adult female is 36 to 53 days, and that of the male is 1 to 3 days.

Nature of Damage

Infestations of the pest remain clustered around the terminal shoots, leaves and fruit, sucking plant sap which results in yellowing, withering and drying of plants and premature shedding of leaves and fruit. Rotting or discoloration can occur where the mealy bugs deposit honeydew. Sometimes rot can start inside the calyx and spread to the interior of the fruit. Sooty mould and wax deposits can block light and air from the plant reducing photosynthesis and hence plant vigour and crop yield.

Management

- ☆ Dead leaves and pruning material should be removed from the field as these may have mealy bugs or eggs on them. Female mealy bugs do not fly or crawl far, so infestations can usually be brought under control in an infested plant. Inspect new plants carefully before putting them in a field; keep them in quarantine for a month.
- ☆ Healthy plants are less susceptible to mealy bug infestations than plants that are weak and stressed. If you only have a few mealy bugs on your plant, try washing them off with a steady stream of water. Repeat if necessary. It may not completely eliminate the problem, but it can help keep a small problem from developing into a big one.
- ☆ A ladybird, *Cryptolaemus montrouzieri* (Mulsant) @ 10/tree can be released to control mealy bugs. A complex of parasites and predators in pomegranate attacks mealy bug. These natural enemies are often capable of providing sufficient control alone or when integrated with chemical treatments. Green lacewings *Chrysoperla carnea* (Stephens) is a generalist predator that feed on mealy bugs in addition to aphids and many other pests, contributing to the control of mealy bugs in pomegranate.
- ☆ Spraying of Methyl-*O*-demeton 25 EC @ 80 ml or Dichlorvos 76 EC @ 70ml/100 litres of water can also give good control. June is good timing, when the crawlers begin moving from the egg masses. Another possible timing is after bud break when most crawlers are at the base of leaves. However, June has the advantage of also being good timing for soft scales.

4. White Fly

White fly *Siphoninus phillyreae* Haliday (Aleyrodidae: Hemiptera) are the tiny yellow insects, which derive their name from the white wax covering on their wings and body. The whitefly is a pest of numerous ornamental and fruit crops. Heavy infestations cause leaf wilt, early leaf drop and smaller fruit.

Host Plants

Pomegranate, pear, apple and loquat.

General Appearance

The adult appears much like a typical whitefly with a light dusting of white wax. Depending on temperature, females live from 15 to 60 days, while males live on an average of 9 days depending upon the temperature. The ability of all life stages to overwinter on non-deciduous hosts allows rapid buildup in population at the start of the season. Winged female lay on an average 140 to 150 eggs on the underside of the leaves. The longevity of female is 19 days. The eggs are pale yellow, covered by a very thin layer of white wax. The eggs are long, elliptical and almost pointed at the front, with a very short stalk. The egg is 220 to 240 µm long and 75 to 85 µm wide. White fly is polyvoltine with several generations per year. Reproduction is sexual and there are four immature instars. When the nymphs emerge, they rarely move far and feed on the plant sap until pupation. Pupal case appears similar to the white armour of a male snow scale. Closer observation with a hand lens reveals a whitefly pupal case with two longitudinal tufts of white wax. Nymphs are short glass like rods of wax along the sides of the body. Both legs and antennae are lost after the first molt and subsequent stages remain fixed to the leaf surface. Adult is powdery white, active during early morning hours. The pest is present at any time of the year, but they tend to reach high numbers in July and August.

Nature of Damage

Nymphs and adults suck the sap from leaves on underside. Whiteflies use their piercing, needlelike mouthparts to suck sap from phloem, the food conducting tissues in plant stems and leaves. Large populations can cause leaves to turn yellow, appear dry, or fall off the plants. Like aphids, whiteflies excrete sugary liquid called honeydew, so leaves may be sticky or covered with black sooty mold that grows on honeydew. The honeydew attracts ants, which interfere with the activities of natural enemies that may control whiteflies and other pests.

Management

- ✰ Field sanitation and removal of alternate host plants. Maintain adequate aeration by proper training and pruning Install yellow sticky traps @ 4 to 10 traps/acre.
- ✰ Release of predator's *viz.*, Coccinellid predator *Cryptolaemus montrouzieri* (Mulsant) and lacewing, *Chrysoperla* spp. (Stephens). Good results have been achieved following the release of the parasitoid *Encarsia inaron* (Walker), *Encarsia haitiensis* (Dozier) and the predator *Clitostethus arcuatus* (Rossi).
- ✰ Thiamethoxam 25 WG @ 0.2 g/litre and Imidacloprid 17.8 SL @ 0.25 ml/litre are most effective in reducing white fly population.

The other insect pests that affect Pomegranate are Shot hole borer, *Scolytus nitidus* (Muller), Thrips, *Scirtothrips dorsalis* (Hood), Stem borer, *Coelosterna scabrator* (Fabricius), Fruit fly, *Bactrocera Zonata* (Saunders), and Bark beetle, *Indarbela quadrinotata* (Walker) are discussed under the insect pests of Apple, Strawberry and Grape.

Chapter 15

Insect Pests of Fig and their Management

The Common Fig, *Ficus carica* Linnaeus is a deciduous tree or shrub in the family Moraceae grown for its edible fruits. The fig tree has numerous spreading branches and palmate leaves. The leaves are deeply lobed and thick with a rough upper surface and hairy lower surface. Fig trees are believed to originate from Western Asia. Figs were first recorded as cultivated plants in southern Arabia in 2900 BC. Figs are either eaten fresh or preserved by drying. Traders and explorers introduced Fig to most countries. It is widely believed that the apple eaten by Adam in the bible was in fact a Fig. Although Fig trees are generally considered to be hardy, they are susceptible to a number of insect pests. More than 50 insect pests are recorded feeding on Fig tree. Some of the major insect pests are as under:

1. Fig Fruit Fly

Fig fruit fly, *Bactrocera dorsalis* Hendel (Tephritidae: Diptera) is native to Africa, and Middle East. The pest is spreading throughout the tropics due to trade and is now found in India. Fruit fly is generally found on damaged and decaying fruit or vegetation. The Fig fruit fly has been found infesting 74 fruit species.

Host Plants

Fig, loquat, apricot, cherry, citrus, persimmon, apple, plum and peach *etc.*

General Appearance

Females lay batches of 3 to 30 eggs in a single fruit. Under optimum conditions, a female can lay more than 3,000 eggs during her lifetime, but under field conditions from 1,200 to 1,500 eggs per female is considered to be the usual production. Apparently, ripe fruits are preferred for oviposition, but immature ones may also be attacked. The pest punctures the skin of the fruit and deposit eggs in few batches depending on the quality of the fruit. Eggs then hatch to larvae in 1 to 3 days and moult twice while feeding on the flesh of the fruit for 9 to 35 days. The mature maggot emerges from the fruit, drops to the ground, and forms a tan to dark brown puparia. The maggot changes into three instars and later into pre pupae. The third instar larvae exits the fruit and burrow into the soil to pupate. In a week or two the adult emerges from the pupae and within a week they became mature and mating continues. Development from egg to adult under summer conditions requires about 16 days. The developmental period may be extended considerably by cool weather. The life cycle takes about 15 to 16 days in summer. Adults live 90 days on an average and feed on honeydew, decaying fruit, plant nectar, bird dung and other substances. The adult is a strong flyer, recorded to travel 30 miles in search of food and sites to lay eggs. This ability allows the fly to infest new areas very quickly.

Nature of Damage

The damage to crops caused by fruit flies result from 1. Oviposition in fruits and other soft tissue of plants. 2. Feeding by the maggot. 3. Decomposition of plant tissue by invading secondary micro-organisms. Maggot feeding in fruits is the most damaging. Damage usually consists of breakdown of tissues and internal rotting associated with maggot infestation, but this varies with the type of fruit attacked. Infested young fruit becomes distorted and usually drop, mature attacked fruits develop water soaked appearance. The maggot's tunnels provide entry points for bacteria and fungi that cause the fruit to rot. When only a few maggots develop, damage consists of an unsightly appearance and reduced marketability because of the egg laying punctures or tissue break down due to the decay.

Management

- ☆ Destroy fallen fruit by burning or boiling before disposal. This will kill pupae and help break the breeding cycle. Wild birds have also been seen digging through infested fallen fruits for maggots. Ants are known to feed on most life stages of fruit flies.
- ☆ Bagging of fruits on the tree with 2 layers of paper bags at 2 to 3 day intervals minimizes fruit fly infestation. The fruits should be bagged at 3 days after anthesis, and the bags should be retained for 5 days for effective control. It is an environmentally safe method for the management of this pest.
- ☆ Hoeing of the tree basins should be done to expose the pupae to their natural enemies.
- ☆ Pheromone traps can monitor Fig fruit flies. These traps attract and kill male flies.
- ☆ Release of parasitoids such as *Opius longicaudatus* (Ashmead) *Opius vandenboschi* (Fullaway), *Opius oophilus* (Fullaway), and *Bracon* spp. (Say) help in managing the Fig fruit fly. Rove beetles, weaver ants, spiders and birds also predate upon fruit flies.
- ☆ Apply Carbofuran 3 G @ 32.5 kg/ha in the soil against the pupal stage of insect.

- ✰ Chemical control of the fruit fly is not so effective. However, insecticides such as Malathion 50 EC @ 0.5ml, Dichlorvos 76 EC @ 1ml and Phosphamidon 40 SL @ 1.25ml/litre of water respectively are moderately effective against the fruit fly.

2. Fig Midge

Fig midge *Anjeerodiplosis peshawarensis* Mani (Cecidomyiidae: Diptera) is a serious pest of Fig, which is its only host. The pest breeds throughout the year, except from April to July. In this period newly formed fruits are not infested, but in other months scarcely 30 per cent, of the fruits remain free from attack. The midges are abundant, hovering under the host tree in the morning and at dusk.

Host Plants

Fig.

General Appearance

The female fly lays minute, oval eggs on one-week-old fruit of the size of a pea. The eggs are laid in cluster of 10 eggs. Incubation takes 3 to 5 days. The entire larval period is passed inside the fruit. There are four larval instars, which differ in size. The entire larval period lasts 3 to 4 weeks. From the middle of April until the July rains, fourth instar larvae remain inside the Figs. Full grown fourth instar larvae bore out of the fruit and drop to the ground. They finally pupate in the soil without forming a cocoon but become covered in a protective case composed of soil particles that adhere to the larval skin. Early fourth instar larvae may also successfully pupate. This period is of 10 to 26 days. It is shortest of 10 days in the rainy season, when both temperature and humidity are high. The pupa is obtect. Mortality in the pupal stage is between 15 and 30 per cent. Emergence of the adults invariably occurs in the early hours of the day. Two or three days before emergence, the protective case becomes detached. Males emerge before females. The ratio of males to females on the day of emergence is 1:2. Copulation starts soon after emergence and lasts for 60 to 90 seconds. There is considerable overlapping of generations. The total duration of life cycle is 5 to 7 weeks. On the basis of the time required for completion of the life cycle in different months, around seven generations are completed in a year.

Nature of Damage

The maggot bores inside the fruit and feeds on the pulp within. The infested fruits become hard, and deformed. The damaged fig ultimately shrivels, withers and drops down prematurely.

Management

- ✰ Sanitation is the most important management practice for the midge. Remove and destroy all dropped and infested fruit on the plant. Collect damaged fruits along with maggots and destroy.
- ✰ Release of parasitoid *Aprostocetus* spp. (Linnaeus) gives good result.

- ☆ Rake up soil to expose pupae and apply Carbofuran 3 G @ 32.5 kg/ha for killing pupae.
- ☆ Only the adult stage of the Fig midge is vulnerable to contact insecticides, because the maggots are protected within the fruit.
- ☆ Spray Dimethoate 30 EC @ 1ml/litre of water or Malathion 50 EC @ 1.4ml/litre of water in August or September when the pest is in severe condition.

3. Fig Tree Borer

Fig tree borer *Phryneta spinator* Fabricius (Cerambycidae: Coleoptera) is indigenous to the west Himalayan ranges and adjoining areas of India and Pakistan. In India the pest is found in the northern states of Jammu and Kashmir, Himachal Pradesh and Uttarakhand. Infestations may lead to yield losses and even to the death of trees. The grubs that initially bore in the trees sub-cortex and later move deeper into the tree cause most damage. Continuous tunneling weakens the wood, causing branches to break or the main stem to collapse. The adults chew on green growing tips and on the bark of twigs.

Host Plants

Fig, pomegranate and walnut.

General Appearance

The female cuts an incision in twigs or in damaged tree bark and places its single eggs into these cuts, laying a total of about 250 eggs during late summer. Eggs take 10 to 18 days to develop. On hatching the grub tunnel into the trunk or branches, the frass falling from these borings collect in crevices below the hole or falls to the ground. Grubs are cannibalistic, so they tend to space themselves out in the wood, avoiding contact with grubs in other tunnels. Grub development often requires more than 6 to 8 months. The grub bore through the sapwood and because of their size, these large tunnels adversely affect foliage and fruit production. Pupation takes place within the trunk, and can make an easy exit when they emerge as adults. The pupal development period lasts 89 to 99 days. The adult form develops inside the pupa and on maturity emerges. The adult beetles emerge in late summer and are nocturnal, may live for several months and can fly for long distances, facilitating their dispersal. The female then needs to mate with a male and find suitable sites on host plants for laying her eggs. The pest has only a single annual generation.

Nature of Damage

Grubs bore into shoot, feed under the bark by cutting circular holes on the bark at 10 to 15 cm distance on branches and finally reach the trunk. Frass comes out from the entrance holes. The affected trees are fragile and often get broken even with moderate wind. The pest remains active from August to November.

Management

- ☆ Cleaning out the entry holes with an iron hook or wire and then plugging the holes with cotton wool soaked in kerosene oil, crude oil or formalin kills the grub. Other methods involve cutting down infested trees, sawing off severely affected branches, and the removal of alternate host plants.

- ☆ The application of Dimethoate 30 EC @ 1ml/litre of water to Fig trees in late summer kill eggs and young grubs. Older grubs hiding in borrows can be killed in situ by the injection of Dichlorvos 76 EC @ 3ml/litre of water and sealing the hole with mud plaster.

4. Fig Blister Mite

Fig blister mite *Aceria ficus* Cotte (Eriophyidae: Acari) is translucent, cigar shaped small phytophagous microscopic mite ranging from 0.15 to 0.3 mm in size that causes deformities on many plants species. These mites are noticed when their feeding causes abnormalities of plant tissues such as gall and erineum formation, leaf curling, blisters, fruit russeting, and deformed buds *etc.* Fortunately, these mites rarely cause serious harm to plants, and control is seldom needed.

Host Plants

Fig.

General Appearance

These mites are colorless to white. Mites generally overwinter as fertilized adult females under bud scales from January to March on protected sites on the host plant, and emerge at bud break in spring. The mites occur first in late May in low density. The density increases rapidly from mid July and have a peak in mid August. Both males and females are present throughout the growing season. Reproduction is continuous. The pest takes 2 to 3 weeks to complete one generation. Around 2 to 3 generations or more are completed in a year.

Nature of Damage

The Fig mite infests bud scales and young leaves. Feeding causes a faint russeting of the leaves, generally in the interior portion of the canopy and may result in leaf drop and stunting of twigs. More importantly, this mite transmits the Fig mosaic virus. The virus is not present in the egg stage of the mite, but once acquired through feeding is retained through molts.

Management

- ☆ To help reduce virus transmission, monitor leaves about a month after they emerge (in May) to detect Fig mites.
- ☆ Application of HMO @ 2 per cent (2 litres/100 litres of water) is effective in controlling overwintering blister mites and should be applied if the mites were a problem the previous year. Sulfur has long been used for mite control. Full coverage is necessary for control.
- ☆ Chemical treatments should be applied before bloom like Fenpyroximate 5 SC or Milbemectin 1 EC @ 100 ml/100 litres of water.

The other insect pests that affect Fig are: San Jose scale, *Quadraspidiotus perniciosus* (Comstock), Chaffer beetle, *Adoretus simplex* (Sharp), Stem borer, *Aeolesthes sarta* (Solsky), Tree borer, *Sanninoidea exitiosa* (Say), Peach twig borer, *Anarsia lineatella* (Zeller), European red mite, *Panonychus ulmi* (Koch) and Shot hole borer, *Scolytus nitidus* (Muller) are discussed under the insect pests of Apple, Plum and Peach.

Chapter 16

Insect Pests of Loquat and their Management

The loquat *Eriobotrya japonica* Thunb native to China and Japan is an evergreen tree known for its small, yellow orange fruits, which are edible. It is also sometimes called Japanese plum, though it is not to be confused with plums, which is another fruit. The tree typically grows to a height of 20 feet, producing dark green, thick leaves and scores of fragrant white blooms. The principle enemies that affect the quality of loquat are as under.

1. Bark Eating Caterpillar

Bark eating caterpillar *Indarbela quadrinotata* Walker (Metarbedelidae: Lepidoptera) is a polyphagous pest causing huge damage to loquat trees. The pest has been reported to infest 70 plant species across fruits, forests and avenue plantations. The caterpillar is nocturnal in nature and feeds on the bark of the host plant under a frass ribbon and remains hidden inside the galleries formed at the forking points on trunk and branches during the day, making its management difficult.

Host Plants

Loquat, plum, mulberry, pomegranate, plum, cherry and apple.

General Appearance

Adult moths are active at the beginning of the wet season in April to July. Female moths deposit eggs below loose bark in clusters of 15 to 25 from April to July in forks in the old wood, or where twigs have broken off or been badly pruned. Eggs hatch in 8 to 10 days after being laid. The larval development is seen in the month of August and the first peak of infestation is also observed during this month. Another peak in larval population is observed during November, which emerges from the eggs laid by the moths late in the month of July. A third peak in larval population is observed during January and February, which coincided with the rainfall during these months. Caterpillars eat the bark and bore inside the tree, feeding for 9 to 10 months. Several caterpillars may attack the same tree at different locations with serious injury to the bark and the death of small branches. Pupal period lasts for 3 to 4 weeks. Moth emergence occurs in the Month of April and November. The holes left on the trunk may lead to infestation by other insects or plant pathogens. Affected trees also break at the points of attack. A severe infestation may arrest the growth of the tree and the fruiting capacity. Total life cycle lasts for around a year. One generation is completed in a year.

Nature of Damage

Caterpillars feed on the bark and make tunnels in the trunk. Due to tunneling, girdling is caused, which result in death of attacked stem. The newly hatched larvae feeds on the bark of the host plant making a dark ribbon like silken web and remains hidden inside the galleries formed at the forking points on trunk and branches during the day, making its management difficult. Blackish larvae can be observed underneath the fresh webbing.

Management

- ✰ Orchard should be kept clean and overcrowding of trees should be avoided. The webs around the affected portion should be cleaned.
- ✰ Inserting an iron spike into each hole to kill the larva is effective when infestation is low. Removal of the frass ribbon alone can reduce the infestation up to 50 per cent.
- ✰ Clear the holes/tunnels with wire followed by injection of kerosene oil or Chlorpyrifos water solution in the ratio of 1:1. Cotton swab soaked in petrol or kerosene should be inserted in the holes and sealed with mud.
- ✰ Application of Phorate 10 CG at the rate of 2 g per hole or to the soil around the collar of the trees @ 50 g/plant gives good control for upto 6 weeks after application.
- ✰ Alternate sprays with Quinalphos 25 EC @ 1 ml/litre of water or Methomyl 40 SP @ 1gm/litre of water or Phosalone 35 EC @ 1.4ml/litre of water is effective in controlling the pest.

2. Green Scale

Coccus viridis Green (Coccidae: Hemiptera) is the major pest of Loquat. It is native to Brazil affecting around 57 plants. It is a soft scale and secretes little wax. Young females are sedentary and little movement is noticed in adults. The adult female is shiny pale green with a conspicuous black marking that is dorsally visible to the naked eye. Adult scales are 2.5 to 3.25 mm. Dead scales are light brown or buff colour and the black internal marking is lost.

Host Plants

Loquat, coffee, guava, avocado, orange and citrus.

General Appearance

Green scale is parthenogenetic and oviparous. A mature female lays whitish oval eggs on the underside of leaves and keeps them underneath her body to protect them. Adults complete egg deposition in 8 to 42 days. Eggs hatch in few hours into crawlers that are flat, oval, yellowish green, moving slowly on the plant or disperse to other hosts. Once a suitable leaf or green shoot is found the nymphs settle and begin to feed. There are three nymphal stages before becoming an adult, each stage being larger than the previous stage. They usually remain in this same spot unless their position becomes unfavourable. The mature female does not move. The undersurface of the leaf is preferred along the midribs and adult scales may be found in a line along both sides of the midrib and lateral leaf veins. Three to four generations are passed in a year.

Nature of Damage

The nymph and adult suck the sap from leaves. The scale insects excrete honeydew on which bees, wasps, ants and other insects feed. The vigour of the plant is reduced. Sooty mould fungus often grows on the honeydew and this decreases the area of leaf available for photosynthesis. The young shoots are often attacked.

Management

- ✰ Prune and destroy the infested shoots at initial stage of infestation. Control ants and dust which can give the scale a competitive advantage.
- ✰ Petroleum oil sprays should be applied @ 2 per cent (2 litres/100 litres of water) in dormant season to control the crawlers.
- ✰ The parasitic wasp *Encarsia* spp. and the predatory beetle *Cryptolaemus montrouzieri* (Mulsant) preys on green scale.
- ✰ The white fringed fungus *Verticillium lecanii* (Zimmerman) @ 10×10^6 spores per ml can cause significant control to the pest.
- ✰ When the pest is in severe form spray Chlorpyrifos 20 EC or Dimethoate 30 EC with dosage of 100 ml/100 litres of water to kill the crawlers and newly settled scale insects.

3. Aphid

Toxoptera aurantii Boyer de Fonscolombe (Aphididae: Hemiptera) the brown citrus aphid, is one of the world's most serious pests of citrus. It is even more of a threat to citrus because of its efficient transmission of citrus tristeza clostero virus (CTV). It also attacks loquat. This aphid is different than all other aphids because there is no sexual stage in the life cycle so the species everywhere are anholocyclic.

Host Plants

Loquat, coffee, mango, litchi, avocado, citrus, orange, *etc.*

General Appearance

The development of this aphid is temperature dependent. Reproduction is partheneogenic or asexual. Eggs are not produced by this species. Females give birth to living young ones. Females start reproducing soon after becoming adults. They produce 5 to 7 live young ones per day, up to a total of about 50 young ones per female. The adult longevity is of around 60 days. Apterous females complete a generation in about 1 to 3 weeks. There are four nymphal stages. The duration of 1st, 2nd, 3rd and 4th nymphal instars are 1.0, 2.0, 2.0, and 1.0 days respectively. They are without wings and brownish in color. Around 30 annual generations are completed in a year.

Nature of Damage

Aphids feed by sucking sap from the plants. This often causes the plants to become deformed, the leaves curled and shriveled. This pest congregates on the tender young shoots, flower buds and the undersides of young leaves. They are not known to feed on the older and tougher plant tissues. It also causes some leaf distortion and malformation in growth of leaves and tips of shoots. In the spring, it is very harmful to fruits, causing flower buds to drop off. The abundant honeydew it produces attracts ants and allows the development of sooty mould.

Management

- Collect and destroy the damaged plant parts. Maintain adequate aeration by proper training and pruning.
- Use yellow sticky traps @ 4 to 10 traps/acre.
- The parasitoid *Aphidius colemani* (Viereck) and predators *Chrysoperla carnea* (Stephens), *Coccinella septempunctata* (Linnaeus) are reported feeding on aphid.
- Spray Chlorpyrifos 20 EC @ 100 ml or Dimethoate 30 EC @ 100ml/100 litres of water after young leaves start to grow. Spray Methyl-*O*-demeton 25 EC @ 80ml/100 litres of water as soon as nymphs are observed.

The other insect pests that affect Loquat are: Fruit fly, *Bactrocera dorsalis* (Hendel), San Jose scale, *Quadraspidiotus perniciosus* (Comstock), Chaffer beetle, *Adoretus simplex* (Sharp), Grey weevil, *Myllocerus discolor* (Boheman), Pomegranate Butterfly, *Deudorix epijarbas* (Moore), Stem borer, *Aeolesthes sarta* (Solsky), Tree borer, *Sanninoidea exitiosa* (Say), European red mite, *Panonychus ulmi* (Koch) and Shot hole borer, *Scolytus nitidus* (Muller) are discussed under the insect pests of apple, plum, peach and fig.

Chapter 17

Insect Pests of Quince and their Management

Quince, *Cydonia oblonga* Mill is a bush like deciduous tree in the family Rosaceae grown for its edible fruits. The quince is the sole member of the genus *Cydonia*. The tree has crowded branches; the leaves have a smooth upper surface and hairy lower surface. The tree produces single white light pink colored flowers on tiny shoots. The tree is self-pollinated. The fruit is initially covered in dense gray white hairs but these disappear as the fruit ripens. The ripe fruit is golden yellow in colour and resembles a pear or apple depending on variety. Quince trees can reach heights of 5 to 8 m and can live for around 50 years, have an economic lifespan of approximately 25 years. Quince originated from Asia Minor. The insect pests attacking quince are as follows:

1. Round Geaded Tree Borer

Round headed tree borer *Saperda candida* Fabricius (Cerambycidae: Coleoptera) is a major pest of quince. All the woody parts of the tree from the buds and twigs to the trunk and roots are susceptible to borer attack. It attacks healthy young trees, boring into trunks and often causing tree death. Once considered a major pest of apples this pest is now rarely a problem in commercial apple orchards.

Host Plants

Quince, apple, cherry, crab apple, plum and pear.

General Appearance

The adult beetle is 1 inch long with the antennae being about the same length as the body. Adult beetles usually appear in May and June. Females lay eggs under the bark, scales or crevices until late July. The female first make egg scars on the bark with their mouthparts. The hatched grub begins feeding within the bark and by

September, the grubs are found between the bark and sapwood. During this time the grub eject woodcuttings and frass from their tunnels. By the time winter sets in, the grub has produced a tunnel about 3 to 4 inches long. The grub passes the winter in the sapwood. During the following spring, summer and fall, the grub bores deeper 1 to 2 inches into the wood. The grub requires two years for development. Pupal stage is of short duration and is the resting stage of the pest. Pupation occurs in the spring of the second year, and adults emerge in early summer. They crawl over the surface of the tree and feed to some extent on the foliage and on the new twig growth. Despite having well developed wings the adult beetle fly short distances. The borer has a two-year life cycle.

Nature of Damage

The grub after hatching feed beneath the bark for a while before entering wood. The presence of frass and woodcuttings on the bark or at the base of the tree and darkened areas in the bark due to sap flow are evidence of attack. One or two borers may kill a young tree, and trees 5 to 10 years old may suddenly break off at the ground because of earlier borer infestations.

Management

- ✰ Destroy all wild trees weeds and grasses. The adults prefer shaded localities for egg laying. Some sort of protection can be obtained by washing the tree trunk with lime sulphur @ 1:7 in the month of May or June.
- ✰ During August and September, check each tree for the presence of the young grub, especially where previous infestations have occurred. The newly hatched grubs cause some sap flow at the point where they begin to feed. This brown sap stain usually can be seen easily on the surface of the bark. A shallow cut in the surface of the bark with a sharp knife will expose or kill the young borer without causing any injury to the tree.
- ✰ The grub can also be killed by a flexible wire probe, which is put into the tunnel.
- ✰ As adult emergence and egg laying can occur over a relatively long period, two to three insecticide applications may be needed during June and July like Chlorpyrifos 20 EC or Dimethoate 30 EC @ 100 ml/100 litres of water.
- ✰ Insert in the holes Celphos @ 1 tablet/hole or Formalin 4 per cent and seal with mud plaster.

2. Quince Curculio

Quince curculio *Conotrachelus crataegi* Walsh (Curculionidae: Coleoptera) is a genus of true weevils that bores into flesh of Quince causing damage and loss of quality of fruit. The size of the adult is 3.5 to 5.8 mm. The pest is native to North America.

Host Plants

Quince, apple, plum, cherry, apricot, crab-apple, avocado, *etc.*

General Appearance

The pest passes winter in the grub stage in soil and change to pupae in spring. The adults appear in summer, in the middle of July. Females lay eggs singly on fruit. They first build an egg chamber under the fruit skin to receive the egg and then turn around and place the egg in the cavity. Then female slices a curved slit underneath the egg cavity, leaving the egg in a flap of flesh and causing a crescent shaped scar on the outside of the fruit. Without this curved slit, eggs are killed by pressure from the growth of the host fruit. After the eggs hatch in 7 to 10 days, later the grub tunnel into the flesh of the fruits, rarely reaching the core. In about one month, usually in August, the mature grub leaves the fruit to make cells 2 to 3 inch deep in the soil where they remain until the next year.

Nature of Damage

The grub and adult cause damage to fruit before maturity. The adult beetle feed on petals, leaves and young fruits. The fruits then fall off prematurely while the grubs develop inside the fruit. The fruit become susceptible to brown rot fungus on the tree.

Management

- ☆ Destroy the fallen and damaged fruit before adult emerges.
- ☆ Biological control using fungi *Beauveria bassiana* (Balsamo), *Metarrhizium anisopliae* (Metchnikoff) and entomopathogenic nematode *Steinernema Heterorhabditis* (Gaugler) reduce the population to some extent.
- ☆ Application of insecticides like Dimethoate 30 EC @ 100 ml or Quinalphos 25 EC @ 100 ml/100 litres of water at pink bud and petal fall stage will control the pest.

The other insect pests which attack Quince are: Oriental fruit moth, *Grapholita molesta* (Busck), Aphid, *Aphis pomi* (DeGeer), Two spotted mite, *Tetranychus urticae* (Koch), Codling Moth, *Cydia pomonella* (Linnaeus), Fruit tree leaf roller, *Archips argyrospila* (Walker), Mealy bug, *Maconellicoccus hirsutus* (Green), Scale insect, *Quadraspidiotus perniciosus* (Comstock) and Round headed apple tree borer, *Saperda candida* (Fabricius) are discussed under the pests of apple, grape and pear.

Chapter 18

Insect Pests of Kiwi Fruit and their Management

Kiwi fruit or Chinese gooseberry, *Actinidia deliciosa* A.Chevalier is a deciduous climbing vine or shrub in the genus Actinidia grown for its edible fruit. The plant is vigorous and woody with nearly heart shaped leaves, which have long petioles and are alternately arranged on the stems. Young leaves and shoots are covered with tiny red hairs, while mature leaves are smooth and dark green on the upper surface and white and downy on lower surface. The kiwi plant produces fragrant white yellow flowers singly or in clusters of 3 at the leaf axils. The fruit is oval in shape with green brown skin covered in stiff brown hairs. The flesh of the fruit is bright green and juicy with many tiny black seeds. Kiwi plants can reach a height of 10 m are mostly insect pollinated, have an economic lifespan of 3 years, after which time fruit production begins to decline. It originated from China and is its national fruit. The insect pests that attack kiwi plant are:

1. Leaf Roller

Leaf rollers are the most important insect species in kiwifruit during the early part of the season. Four leaf rolling lepidopterous insect pests damage the kiwifruit. They are: omnivorous leaf roller *Platynota stultana* (Walsingham), fruit tree leaf roller *Archips argyrospila* (Walker), oblique-banded leaf roller, *Choristoneura rosaceana* (Harris) and orange tortrix *Argyrotaenia citrana* (Fernald) belonging to family Tortricidae and order Lepidoptera. Among all, the omnivorous leaf roller *Platynota stultana* is the most damaging. This troublesome species has many hosts, and populations can become quite high if not managed. Natural control of leaf rollers can occur, but generally will not reduce populations below damaging level.

Host Plants

Kiwi, grape, peach, pear, plum, walnut and pPomegranate.

General Appearance

Omnivorous leaf roller caterpillars may differ in body colour from cream to brown. Omnivorous leaf roller has four to six generations per year depending on climatic conditions. Eggs are laid on foliage in clusters or masses containing an average of 97 individual eggs per mass in one night and total of 500 to 600 eggs per female in 5 to 6 days. Eggs are covered with sticky adhesive material. Eggs hatch in 5 to 9 days. There are five larval instars. Newly hatched larvae move towards the top of the plant and feed within a bud or between two leaves. Later instars feed within a shelter constructed of rolled or folded leaves. Larval period is of 20 to 30 days. Fifth instar larvae overwinter in webbed nests. Pupation takes place in a rolled leaf. Pupal stage requires 4 to 9 days.

Nature of Damage

Omnivorous leaf roller and the other leaf rolling caterpillars directly damage fruit by scarring the surface when they feed. They sometimes make holes where two or more fruits touch each other. Leaf rolling caterpillars can also feed on developing buds and flowers. The caterpillars tie leaves around themselves with silken threads and eat the leaves from within this shelter.

Management

- ☆ Remove mummified fruit from trees in winter. Early harvest can also prevent infestation by fourth generation larvae.
- ☆ Various parasitic wasps and tachinid flies attack leaf roller eggs, larvae or pupae.
- ☆ Applications of *Bacillus thuringiensis* are acceptable in an organically certified crop.
- ☆ Mating disruption pheromones are very effective as stand alone products for reducing infestations and plant damage by omnivorous leaf roller.
- ☆ Spray Quinalphos 25 EC @ 100 ml or Phosalone 35 EC @ 140 ml/100 litres of water when pest population is high.

The other insect pests that attacks Kiwi are: Two spotted mite, *Tetranychus urticae* (Koch), Scale insect, *Quadraspidiotus perniciosus* (Comstock) and Thrips, *Thrips* spp., are discussed under the insect pests of Apple.

Chapter 19

Insect Pests of Persimmon and their Management

Persimmon *Diospyros kaki* Thunb (Ebenaceae: Ericales) is the edible fruit and the most widely cultivated in the temperate regions of the world. The most widely cultivated species is the Japanese persimmon. The colour of the ripe fruit ranges from yellow orange depending on the species and variety. The calyx generally remains attached to the fruit after harvesting, but becomes easy to remove once the fruit is ripe. Persimmon plants form a low head and develop a framework of strong branches. In general, cultivars differ markedly in vigour and growth habit. Some cultivars are dwarf and highly precocious, whereas others are vigorous, upright and late maturing. Generally persimmons have few insect pests, which are listed as under:

1. Long Tailed Mealy Bug

Tailed mealy bug *Pseudococcus longispinus* Targioni-Tozzetti (Pseudococcidae: Hemiptera) gets its name from the two long waxy filaments on the last abdominal segment of female and differentiates it from other mealy bugs.

Host Plants

Persimmon, grape, peach, plum, apple, pear, *etc.*

General Appearance

The long tailed mealy bug reproduces sexually; each female lays 75 to 200 eggs over a 10 to 20 day period. It is believed that mealy bug has no visible egg stage and the nymphs hatch immediately after oviposition. The nymph is similar to the larger adult female except that the filaments around the edges are shorter. The nymphs undergo three instars in females and four in males. The body is covered by white waxy dust but first instar lacks it. Females grow bigger in size than males. The males feed only during first and second instar. Third instar is the pre pupal stage and migrates to the protected place where they secrete waxy cocoon in which they complete development. Pupating and adult males do not feed. Females feed in all instars. The nymphal stage lasts for 22 to 24 days. Females live for 2 to 3 months and males only a few days. Mealy bugs complete 2 to 6 generations a year.

Nature of Damage

The pest feed by sucking out plant sap from leaves and stems. Honeydew and sooty mold further disfigure infested plants, which may eventually be killed. The pests also secrete a fluffy white wax, which signifies the infested plants. High population feeding causes slow growth and immature leaf drop.

Management

- ☆ Removal of weeds and alternate host plants in and nearby orchard throughout the year.
- ☆ Remove and destroy the loose bark.
- ☆ Do not over water or over fertilize because mealy bugs are attracted to plants with high nitrogen levels and soft growth.
- ☆ The exotic predator, *Cryptolaemus montrouzieri* (Mulsant) controls the pest if infestation is low.
- ☆ Spray of HMO's at dormant stage @ 2 per cent (2 litres in 100 litres of water) provides suppression of mealy bugs.
- ☆ Aerial spraying of insecticides like Methyl-*O*-demeton 25 EC @ 80 ml, or Dimethoate 30 EC @ 100 ml or Chlorpyrifos 20 EC @ 100 ml/100 litres of water can be sprayed to persimmon trees to control various stages of mealy bugs.

2. European Fruit Scale

European fruit scale *Parthenolecanium corni* Bouche (Coccidae: Hemiptera) also called European fruit Lecanium scale or brown apricot scale is a soft scale of shade trees which infests leaves, twigs and fruits and produces honeydew which promotes growth of sooty mould. The pest reproduces parthenogenetically.

Host Plants

Persimmon, grape, apricot, plum, apple, raspberry, *etc.*

General Appearance

The pest overwinters under the bark. Overwintering nymphs mature into adults in April. They mate and female scale lay about 1000 or more eggs during May for a period of 30 days and populations can build quickly. Males die within a few days after mating. Crawlers hatch from early June to mid July. Crawlers move to the leaves where they settle near leaf veins. In late August, crawlers move back to the twigs. Newly hatched crawlers are initially white and become yellow with age. The nymphal stage lasts for 30 to 60 days. There is single generation of scale insect in a year.

Nature of Damage

The pest sucks the sap from branches, fruits and leaves resulting in devitalization of tree, reduced growth, poor fruit size and dead limbs. The most common damage is due to honeydew, which is excreted by feeding nymphs.

Management

- ☆ Prune heavily infested plant parts and destroy them by burning.
- ☆ Minute pirate bugs, lacewings, ladybird beetles, predaceous midges and parasitoid wasps feed on these scale insects.
- ☆ Dormant oil spray of HMO @ 2 per cent (2 litres in 100 litres of water) provides suppression of scale insects especially overwintering nymphs which are on the barks. Spraying should be done when weather is clear and temperature is above 4°C.
- ☆ Spray Chlorpyrifos 20 EC @ 100 ml or Dimethoate 30 EC @ 100 ml in 100 litres of water during summer to kill the crawlers and newly settled scale insects.

The other insect pests that affect Persimmon are: Two spotted mite, *Tetranychus urticae* (Koch), Fruit fly, *Bactrocera dorsalis* (Hendel) and Leaf roller, *Archips argyrospila* (Walker) are discussed under the pests of apple and fig.

Bibliography

Abbasi, M. A., Raza, A., Riaz, T., Shahzadi, T., Rehman, J. M., Shahwar, D, Siddiqui, S.Z., Chaudhary, A.R. Ahmad, N. 2010. Investigation on the volatile constituents of Juglans regia and their in vitro antioxidant potential. *Pakistan Academic Science.*, 47, 137-141.

Abd-Rabou, S. 1998. The efficacy of indigenous parasitoids in the biological control of *Siphoninus phillyreae* (Homoptera: Aleyrodidae) on pomegranate in Egypt. *Pan-Pacific Entomologist* 74: 169–173.

Abd- Rabou, S. 2006. Biological control of the Pomegranate Whitefly, *Siphoninus phillyreae* (Homoptera: Aleyrodidae:) by using the bioagent, *Clitostethus arcuatus* (Coleoptera: Coccinellidae). *Journal of Entomology.*, 3(4): 331-335.

Abdel khalek.A. A., 1993. Studies on mites associated with fig trees in Qaliubia governorate. M.Sc. Thesis, Faculty of Science, *Ain Shams University*. 143 pp.

Abdel Rahman, F. and Maggenti, A. R. 1988. *Gracilacus elongata* sp. (Nemata:. Criconematioidea) parasitic on *Juncus ensifolius* Wikstr. from Mendocino California. *Revue de Nématologie,* 11 (3), 303–306.

Adams, R.E. and Eichenmuler, J.J. 1962. *Gracilacus capitatus* from scarlet oak in West Virginia. *Nematologica,* 8 (2), 87–92.

Agarwal M. L. and Kapoor V. C. 1982. Acanthiophilus helianthi (Rossi) and Chaetostoma completum Kapoor etal. (Diptera: Tephritidae), a serious pests of *Centaurea cyanus* Linnaeus (Compositae) in India. *Journal of Entomological Research* 6(1) 102-106.

Agarwal ML and Kapoor V. C. 1985. On a collection of Trypetinae (Diptera: Tephritidae) from Northern India. *Annals of Entomology*.3: 59-64.

Agarwal ML and Kapoor V.C. 1988. Four new species of fruit flies (Diptera: Tephritidae: Tephritini), together with description and their distribution. *Journal of Entomological Research*.12 (2): 117-188.

Agarwal ML and Sueyoshi Masahiro 2005. Catalogue of Indian fruit flies (Diptera; Tephritidae). *Oriental Insects*. 39: 371-433.

Agarwal ML, Sharma DD, Rahman.O.1987. Melon fruit fly and its control. *Indian Horticulture*; 32: 10–11.

Agata, Z., Filipescu, C., Georgeseu, T., Talmaciu, N., and Bernardis, R. 2007. Biology, Ecology and integrated control of the species *Lyonetia clerkella*: Pest in the apple plantation from Neamt County. *Neamt County Phytosanitary Board*. pp: 1125-1128.

Ahmed, D., and Bhat, M. R. 1987. Insect pests of apple trees in Kashmir. *Geobios new Reports*, 6: 60-63.

Allen, M.W. and Jensen, H.J. 1950. *Cacopaurus epacris*, new species (Nematoda: Criconematidae), a nematode parasite of California black walnut roots. *Proceedings of the Helminthological Society of Washington*, 17 (1), 10–14.

Ananda, N.,Kotical,Y.K. and Balikai,R.A. 2009. Management practices for major sucking pests of pomegranate. *Karnataka Journal of Agricultural Sciences*.,22 (4):790-795.

Anderson, R.V. and Kimpinski, J. 1977. *Paratylenchus labiosus* n. sp. (Nematoda: Paratylenchidae) from Canada. *Canadian Journal of Zoology*, 55, 1992–1996.

Andrássy, I. (2007) Free-living nematodes of Hungary, II (*Nematoda errantia*). *Hungarian Natural History Museum, Budapest*, 496 pp.

Annis B, Tamaki G, Berry RE. 1981. Seasonal occurrence of wild secondary hosts of the green peach aphid, *Myzus persicae* (Sulzer), in agricultural systems in the Yakima Valley. *Environmental Entomology*.10: 307-312.

Anonymous 1983. Manual for pesticides users. New Delhi, India: Pesticides Association of India. 191 pp.

Anonymous, 2008. *Statistical Report*. Directorate of Horticulture, J&K Rajbagh Srinagar, pp. 55-75.

Anonymous 2012. Pest Risk Assessment of *Drosophila suzukii:* Spotted-wing drosophila (Diptera: Drosophilidae) on fresh fruit from U.S.A. Final MDI Technical Paper No. 2012/05, New Zealand Government.

Armstrong, J.W.; Hu, B.K.S.; Brown, S.A. 1995. Single-temperature forced hot-air quarantine treatment to control fruit flies (Diptera: Tephritidae) in papaya. *Journal of Economic Entomology*. 88, 678-682.

Arshad A, Wani N. A, Ahmad S. B, Wani A. R., and Munib M. 2015. Incidence and Relative Bio-Efficacy of Different Insecticides against Chaetoprocta (*Chaetoprocta odata* Hewitson) Infesting Walnut in Kashmir Valley: *Journal of Agricultural Sciences*.7 (9)

Arthurs. S, McKenzie CL, Jianjun C, Dogramaci M, Brennan M, Houben K, Osborne L. 2009. Evaluation of *Neoseiulus cucumeris* and *Amblyseius swirskii* (Acari: Phytoseiidae) as biological control agents of chilli thrips, *Scirtothrips dorsalis* (Thysanoptera: Thripidae) on pepper. *Biological Control* 49: 91-96.

Attia,A. R. 2006. Biological control of the striped mealybug, *Ferrisia virgata* (Ckll.) (Homoptera: Pseudococcidae) on the mulberry tree, *Morus alba* using the coccinellid predator, *Scymnus syriacus* Mars. *Egyptian Journal of Biological Pest Control* 16: 45-50.

Atwal, A. S. and Dhaliwal, G. S. 1999. Agricultural Pests of South Asia and their management. Kalyani publishers. pp 1-484.

Bajaj, H.K. 1988. On the species of *Paratylenchus Micoletzky* (Nematoda: Criconematina) from Haryana, India. *Indian Journal of Nematology*, 17, 318–326.

Balikai RA and Bagali AN 2000. Population density of mealybug, *Maconellicoccus hirsutus* on ber (*Zizyphus mauritiana*) and economic losses. *Agricultural Science Digest* **20**, 62–63.

Banks CJ, Macaulay EDM, Holman J. 1968. Cannibalism and predation by aphids. *Nature*;218: 491.

Barnes MM, Sibbet GS. 1971. Walnut aphid effects on walnut production and quality. *California Agriculture*; 5(1):12-15.

Barnes, M.M., Davis, C.S., Sibbett, G. S. and Barnett, W.W. 1978. Integrated pest management *Journal of pest science*; 88 (.4): 753-765.

Barnes,M.M. 1991. Codling moth occurrence, host race formation and damage, In; vander Geest, L.P.S., Evenhuis,H.H.(Eds.), World crop pests, Vol. 5:Tortricid Pests. Their biology, Natural enemies and control.*Elsevier Amesterdam*: 313-327.

Barrowclough, G. F. 1992. Systematics, biodiversity, and conservation biology, *Columbia University Press, New York*: 220 pp.

Bateman, M.A. 1982. Chemical methods for suppression or eradication of fruit fly populations. In: Economic fruit flies of the South Pacific Region, Queensland Department of Primary Industries, Brisbane, Australia. pp. 115-128.

Bayhan, E., Ölmez-Bayhan, S., Ulusoy, M.R. and Brown, J.K. 2005. Effect of temperature on the biology of *Aphis punicae* (Passerini) (Homoptera: Aphididae) on pomegranate. *Environmental Entomology* 34: 22-26.

Beeson, C. F. C. 1941. The ecology and control of forest insects of India and their neighbouring countries, *Vasant press: Dehradun*: 785.

Beirne, B.P. 1972. Pest insects of annual crop plants in Canada. IV. Hemiptera-Homoptera V. Orthoptera VI. Other groups. *Memoirs of the Entomological Society of Canada* 85. 73 pp.

Bellows, T.S, Paine T.D, Arakawa KY, Meisenbacher C, Leddy P, Kabashimo J. 1990. Biological control sought for ash whitefly. *California Agriculture*; 44: 4-6.

Bess, H. A. 1958. The Green Scale, *Coccus viridis* (green) (Homoptera: Coccidae), and Ants. *Proceeding of Hawaiian Entomological Society*. 16(3): 349-355.

Bezzi, M 1913. Indian trypaneids (fruit flies) in the collection of Indian Museum, Calcutta. *Memoirs of Indian Museum*;3: 53-175.

Bhagat, R. C and Lone M.A 1984. New records and host range of predators of aphids (Aphididae: Homoptera) in Kashmir Valley, India. *Science and Culture*: 50(12) 364-366.

Bhat, D. M, Bhagat, R. C and Aijaz, Q. 2011. A survey of insect pests damaging vegetable crops in Kashmir valley (India), with some new records. *Journal of Entomological Research*; 35(1): 85-91.

Bhat, R. M. 1991. Distribution and Host range of insect pests in Kashmir. *Geobios New Reports*;10: 160-161.

Bishop, G. W, J.W. Guthrie. 1964. Home gardens as a source of the green peach aphid and virus diseases in Idaho. *American Potato Journal*; 41: 28-34.

Bishop, G.W. 1965. Green peach aphid distribution and potato leaf roll virus occurrence in the seed potato producing areas of Idaho. *Journal of Economic Entomology*; 58: 150-153.

Blackman RL, Eastop VF. 1984.Aphids on the World's Crops: An *Identification and Information Guide.* John Wiley and Sons, Chichester, England. 466 pp.

Blatchley, W.S. and C.W. Leng. 1916. Rhynchophora or weevils of North Eastern America. The Nature Publishing Company, *Indianapolis*. 682 pp.

Bose, M., 1943. Bionomics and life history of *Myllocerus laetivirens* Mshl. (Otiorrhynchinae: Curculionidae). *Indian Journal of Entomology*. 5: 103-105.

Brunetti 1917. Diptera of the Simla Distict. *Record Indian Museum* 13: 59-101.

Brzeski, M.W. 1976.*Paratylenchus bukowinensis*, descriptions of plant-parasitic nematodes. Set 6, N 79. Commonwealth Agricultural Bureaux, *Farnham Royal, UK*, 2 pp.

Buhroo, A. A. and Lakatos, F. 2007. On the Biology of the Bark Beetle *Scolytus nitidus* Schedl (Coleoptera: Scolytidae) attacking apple orchards. *Acta Silvatica et Lignaria Hungarica*. 3: 65-74.

Burikam, I.; Sarnthoy, P.; Charensom, K.; Kanno, T.; Homma, H. 1992. Cold temperature treatment for mangosteens infested with the oriental fruit fly. *Journal of Economic Entomology*; 85: 2298- 2301.

Butani, D. K. 1979. Insects and fruits : *Periodical Expert Book Agency*, pp. 259-261.

Butler, B., Lankford, D. Fall 2002. Bearing Red Raspberry Production in Maryland Tunnels. *The Bramble*, 18 (2).

Cannella, C., and Dernini, S. 2004. Walnut: Insights and Nutrition Value. *Proceedings of Vth International Symposium*- Sorrento (Italy) November 9-13: 549.

Capinera JL. 2001. Handbook of Vegetable Pests. Academic Press, *San Diego*. 729 pp.

Carrillo, R., C. Cifuentes, and M. Neira. 2001. Seasonal cycle of *Parthenolecanium corni* (Bouche) (Hemiptera, Coccidae) on *Ribes* spp. in southern Chile. *Agro Sur* 29: 110-113.

Carver, M. 1978. The black citrus aphids, *Toxoptera citricidus* (Kirkaldy) and *T. aurantii* (Boyer de Fonscolombe) (Homoptera: Aphididae). *Journal of Australian Entomological Society.* 17: 263-270.

Charanasri, V. and T. Nishida. 1975. Relative Abundance of Three Coccinellid Predators of the Green Scale, *Coccus viridis* (Green) on Plumeria Trees. *Proceeding of Hawaiian Entomological Society;* 22(1): 27-32.

Choudhuri, J. C. B. 1963. New host plant of *Coelosterna scrabrator* Fabr. (Order: Coleoptera). *Current Science,* **9**: 434-435.

Christenson, L.D.; Foote, R.H. 1960. Biology of fruit flies. *Annual Review of Entomology;* 5, 171- 192.

Ciampolini, M. and Trematerra, P. 1992. Widespread occurrence of walnut fly (*Rhagoletis completa* Cresson) in northern Italy. *Informatore Agrario* **48** : 52-56.

Clarke, R.G.; Howitt, A.J. 1975. Development of the strawberry weevil under laboratory and field conditions. *Annals of the entomological society of America.,*68, 715-718.

Clausen, C. P. (Ed.), B. R. Bartlett, E. C. Bay, P. DeBach, R. D. Goeden, E. F. Legner, J. A. McMurtry, E. R. Oatman, and D. Rosen.1978. Green Scale, (*Coccus viridis* (Green)). In Introduced Parasites and Predators of Arthropod Pests and Weeds: A World View. *Agriculture Handbook No. 480. US Department of Agriculture. Washington,* DC. pp. 73-74

Clausen, C.P.; Clancy, D.W.; Chock, Q.C. 1965. Biological control of the oriental fruit fly (*Dacus dorsalis* Hendel) and other fruit flies in Hawaii. United States Department of Agriculture, *Technical Bulletin No.* 1322, 102 pp.

Copland, M. J. W. and A. G. Ibrahim. 1985. Biology of Glasshouse Scale Insects and Their Parasitoids. In: Biological Pest Control. The Glasshouse Experience. *Cornell University Press; Ithaca, New York.* pp. 87-90.

Cottier W. 1953. Aphids of New Zealand. *New Zealand Department of Scientific and Industrial Research Bulletin;*106. 382 pp.

Cross, J.V., Solomon, M. G., Babandreier, D., Blommers, L., Easterbrook, M. A., Jay, C. N., Jenser, G., Jolly, R. L., Kuhlmann, U., Lilley, R., Olivella, E., Toepfer, S., and Vidal,S. 1999. Biocontrol of pests of apples and pears in northern and central Europe: *Biocontrol Science and Technology,* 9, 277–314.

Cunningham, R.T. 1989. Biology and physiology; Para pheromones. Fruit flies; their biology, natural enemies and control. pp. 221-230. Elsevier, Amsterdam, Netherlands.

Curtis,C.E.1976. Economics of control and implenting orchard cleanup.*Almond* facts, 41:5-8.

Dar, G. A., Sheikh, A. G., and Ganjoo, B. L. 1977. Relative efficacy of some insecticides in suppressing *Lymantria obfuscata* walker on apple trees in Kashmir. *Pesticides,*11(10), 27-29.

David, B.V. 1992. Pest management and pesticides. *Indian Scenario*;Namrutha Publications, Madras. 384 pp.

Davidson, R. H. and William, F. I. 1987. Insect pests of farm, garden and orchard 8: 413-447.

Dawson GW, Griffiths DC, Merritt LA, Mudd A, Pickett JA, Wadhams LJ, Woodcock CM. (1990).Aphid semiochemicals- a review, and recent advances on the sex pheromone. *Journal of Chemical Ecology* 16: 3019-3030.

Dekle, G. W. 1976. Green Scale, *Coccus viridis* (Green). Div. of Plant Industry. *Entomology Circular No*. 165.

Dennill, G.B. 1991. A pruning technique for saving vineyards severely infested by the grape bud mite *Colomerus vitis* (Pagenstecher) (Eriophyidae). *Crop Protection* **10**: 310-314.

Dhiman, S.C and Arif, M. 2006. Insect pests of temperate fruit crops and their management, Indus publishing Co., New Delhi, 3:280-292.

Dolinski, C.M., Souza, R.M. and Huang, S.P. 1996. *Paratylenchus emmoti* sp. (Nemata: Tylenchulidae) found in the Cerrado of central Brazil. *Fitopatologia Brasileira,* 21 (4), 414–417.

Doucet, M.E. 1994. New data on *Gracilacus colina* (Nemata: Criconematoidea). *Fundamental and applied Nematology*, 17 (2), 117–121.

Dreistadt SH, Flint ML, 1995. Ash whitefly (Homoptera: Aleyrodidae) overwintering and biological control by *Encarsia inaron* (Hymenoptera: Aphelinidae) in northern California. *Environmental Entomology*, 24(2): 459-464.

Drew, R.A.I. and Raghu, S. 2002. The fruit fly fauna (Diptera: Tephritidae) of the rain forest habitat of the Western Ghats, India. *The Raffles Bulletin of Zoology* 50(2):327-352.

Drew, R.A.I. 1982. Economic fruit flies of the South Pacific (2nd edition), Queensland Department of Primary Industries, Brisbane, Australia. Drew, pp. 129-139.

Drooz, A. T. 1985. Insects of eastern forests. USDA *Miscellaneous Publication* :1426

Duso,C. and G. Dal lao. 2006. Life cycle, phenology and economic importance of the walnut husk fly *Rhagoletis Completa* Cresson (Diptera:Tephritidae) in northern Italy. *Annales de la societe Entomologique de france* 42(2): 245-254.

Dusso, C. and De Lillo, E. 1996. Grape. Eriophyoid Mites, Their Biology, Natural Enemies and Control (Ed. by Lindquist, E.E., Sabelis, M.W. and Bruin, J.), Elsevier, Amsterdam. pp. 571-582.

Eayre, C., Sims, J., Ohr, H., and Mackey, B. 2000. Evaluation of methyl iodide for control of peach replant disorder. *Plant Disease*; 84:1177-1179.

Edward, J.C. and Misra, S.L. 1963. *Paratylenchus nainianus* sp. (Nematoda: Criconematidae) from Uttar Pradesh, India. *Nematologica*, 9 (2), 215–217.

Elkinton, J.S. and Liebhold, A. M. 1990. Population dynamics of gypsy moth in North America. *Annual Review of Entomology* **35**(4): 571-596.

Erlandson, W., and Obrycki, J. 2015. Population Dynamics of *Empoasca fabae* (Hemiptera: Cicadellidae) in Central Iowa Alfalfa Fields. *Journal of Insect Science* **15**(1): 1-6.

Faeth, S. H., Mopper, S., and Simberloff, D. 1981. Abundances and diversity of leaf-mining insects on three oak host species: effects of host- plant phenology and nitrogen content of leaves. *Oikos*, 37: 238-251.

Faostat, 2011. Faostat Database Results 2009.(accessed January, 2011).

Ferro.D.N, Mac.Kenzie JD, Margolies DC. 1980. Effect of mineral oil and a systemic insecticide on field spread of aphid-borne maize dwarf mosaic virus in sweet corn. *Journal of Economic Entomology*; 73: 730-735.

Fink DE. 1932. Biology and habits of the strawberry leafroller, *Ancylis comptana* (Froel.), in New Jersey. *Journal of Agricultural Research*;44: 541-558.

Flanders KL, Radcliffe EB, Ragsdale DW. 1991. Potato leaf roll virus spread in relation to densities of green peach aphid (Homoptera: Aphididae): implications for management thresholds for Minnesota seed potatoes. *Journal of Economic Entomology*; 84: 1028-1036.

Fletcher, B.S. 1989. Ecology; movements of tephritid fruit flies, their biology, natural enemies and control (Ed. by Robinson, A.S.; Hooper, G.), Elsevier, Amsterdam, Netherlands. pp. 209- 219.

Fredrick JM. 1943. Some preliminary investigations of the green scale, *Coccus viridis* (Green), in south Florida. *Florida Entomologist* ; 26: 12-15; 25-29.

Fuester, R. W., Swan, K. S., Taylor, P. B. and Ramasesiah, G. 2008. Effects of parent age at mating on reproductive response of *Glyptapanteles flavocoxis*, a larval parasitoid of the gypsy moth. *Journal of Economic Entomology*, 101 (4): 1140-1145.

Gaffar,S.A and A.A.Bhat. 1991. Management of Stem borer, *Aeolesthes sarta*, infesting walnut trees in Kashmir. *Indian journal of for*estry 14:138 -141.

Gandev, S. 2007. Budding and grafting of the walnut (*Juglans regia* L.) and their effectiveness in Bulgaria (Review). Bulgar. *Journal of Agricultural Sciences* 13: 683-689.

Ganie SA, Khan ZH and Padder SA (2013b). Identification and taxonomical studies of fruit flies on cucurbits in Kashmir valley. *The Bioscan* 8(1) 263-269.

Ganie SA, Khan ZH, Ahangar RA, Bhat HA and Hussain B (2013a). Population dynamics, distribution and species diversity of fruit flies on cucurbits in Kashmir valley India. *Journal of Insect Science*;13(65) 1-7.

Ghose SK 1971. Morphology of various instars of both sexes of the mealy bug, *Maconellicoccus hirsutus*. *Indian Journal of Agricultural Sciences*; 41:602– 611.

Ghose SK 1972. Biology of the mealy bug, *Maconellicoccus hirsutus* (Green) (Pseudococcidae: Hemiptera). *Indian Agriculture*; 16: 323-332.

Gibson RW, Pickett JA, Dawson GW, Rice AD, Stribley MF. 1984. Effects of aphid alarm pheromone derivatives and related compounds on non- and semi- persistent plant virus transmission by *Myzus persicae*. *Annals of Applied Biology*;104: 203-209.

Gibson, K. E. and Kearby, W. H. 1978. Seasonal life history of the walnut husk flies and husk maggot in Missouri. *Environmental Entomology* **7**(1): 81-87.

Gilkeson LA, Hill SB. 1987. Release rates for control of green peach aphid (Homoptera: Aphididae) by the predatory midge *Aphidoletes aphidimyza* (Diptera: Cecidomyiidae) under winter greenhouse conditions. *Journal of Economic Entomology*; 80: 147-150.

Gill, R. J. 1988. The scale insects of California, (Homoptera: Coccidae). *Technical Services in Agricultural Biosystematics and Plant Pathology*, California Department of Food and Agriculture, Sacramento, CA.

Gilmore JE, 1960. Biology of the black cherry aphid in the Willamette Valley, Oregon. *Journal of Economic Entomology*, 53:659-661.

Gilmore, J.E. 1989. Fruit flies; their biology, natural enemies and control (Ed. by Robinson, A.S.; Hooper, G.), *Elsevier, Amsterdam, Netherlands*. pp. 353-363.

Goeden, R.D. and D.M. Norris. 1965. Some biological and ecological aspects of ovipositional attack in *Carya spp*. by *Scolytus quadrispinosus* (Coleoptera: Scolytidae). *Annals of the Entomological Society of America*; 58:771-777.

Goolsby JA, Kirk AA, Meyerdirk DE 2002. Seasonal phenology and natural enemies of *Maconellicoccus hirsutus* (Hemiptera: Pseudococcidae) in Australia. *Fla. Entomol*. 85: 494-498.

Gries, R., Schaefer, P. W., Hahn, R., Khaskin, G., Gujjandadu, R., Singh, B., Hehar, G. K. and Gries, G. 2007. Sex pheromone component of Indian gypsy moth, *Lymantria obfuscata* Walker. *Journal of Chemical Ecology*, 33 (9): 1774-1786.

Gul.Shaheen, 2011. Biology and management of mealy bug *Drosicha dalbergiae* (stebbing) (Homoptera: Margarodidae) on Almond. (*Prunus dulcis* Mill) Ph.D thesis submitted to SKUAST-Kashmir. 123 pp.

Gunn, D. 1919. The fig and willow borer (*Phryneta spinator*). Union of South Africa, Department of Agriculture, *Bulletin No. 6*: 1-22.

Gupta.S.K 1985. Handbook on Plant mites of India. *Sri aurobindo press,* Calcutta. pp 39-279.

H. F. Barnes 1930. On the biology of the gall-midges (Cecidomyidae) attacking meadow foxtail grass (*Alopecurus pratensis*), including the description of one new species, *Annals of applied Biology;* 17: 339–366.

H. F. Barnes 1936. Insect fluctuations: population studies in the gall midges (Cecidomyiidae *Annals of applied Biology;* 23 : 433–440.

Haliday AH. 1835. *Aleyrodes phillyreae. Entomology Magazine* 2: 119-120.

Hampson, G. F. 1892. Fauna of British India, including Ceylon and Burma moths. *Indian Forest Record*, 1 (23): 527.

Hansen, E.A., J.E. Funderburk, S.R. Reitz, S. Ramachandran, J.E. Eger, and H. McAuslane. 2003. Within-plant distribution of *Frankliniella* species (Thysanoptera: Thripidae) and *Orius insidiosus* (Heteroptera: Anthocoridae) in field pepper. *Environmental Entomology*. 32(5): 1035-1044.

Hardy, D.E. 1949. Studies in Hawaiian fruit flies. *Proceedings of the Entomological Society of Washington*; 51: 181-205.

Haymer, D.S, Tanaka, T. Teramae, C. 1994. DNA probes can be used to discriminate between tephritid species at all stages of the life cycle. *Journal of Economic Entomology* 87:741-746.

Heathcote GD. 1962. The suitability of some plant hosts for the development of the peach-potato aphid, *Myzus persicae* (Sulzer). *Entomologica Experimentalis et Appliciata* 5: 114-118.

Heshamul Heque and Miss Qamar Rahim Malik 1967. Control of fruit flies *DacusZonatus* Saunders by gamma-rays. *International Journal of Applied Radiation and Isotopes*. 18(9): 658.

Highland, H.A. 1956. The biology of *Ferrisiana virgata*, a pest of azaleas. *Journal of Economic Entomology* **49**: 276-277.

Hoffman, G., D, Hogg, and Boush, M. 1991. Potato Leafhopper (Homoptera: Cicadellidae) Life History Traits on Water-Stressed Alfalfa in the Early Regrowth and Bud Stage. *Environmental Entomology* **20**(4): 1058-1066.

Hollingsworth CS, Gatsonis CA. 1990. Sequential sampling plans for green peach aphid (Homoptera: Aphididae) on potato. *Journal of Economic Entomology*; 83: 1365-1369.

Horsfall J. L 1924. Life history studies of *Myzus persicae* Sulzer. *Pennsylvania Agric. Agricultural Experiment Station Bulletin*;185:16.

Houser, J. S. 1918. Insect pests of Ohio shade and forest trees. *Ohio Agricultural Experiment Station Bulletin*: 332.

Hussain, M. A. and Khan, A. W.1949. Bionomics and control of walnut weevil (Alcidesporrectirostris Marshall). *Indian Journal of Entomology*;11: 77-82.

Jagginavar, S. B., N. D. Sunitha and D. R. Paitl 2008. Management strategies for grape stem borer *Celosterna scrabrator* Fabr. (Coleoptera: Cerambycidae) *Indian Journal of Agricultural Research*, 42(4): 307-309.

Janjua, N. A., Mustafa, A. M. and Samual, C. K. 1943. On the biology and control of codling moth *Cydia pomonella* L., in Baluchistan. *Indian Journal of Agricultural Sciences*;13: 112-128.

Janjua, N.A. 1938. Codling moth in Afghanistan, *Current Science*; 7: 125.

Janjua, N.A. and Samuel, C. K. (1941). Fruit pests of Baluchistan. *ICAR miscellaneous bulletin No*. 42: 1-41.

Jansson RK, Smilowitz Z. 1986. Influence of nitrogen on population parameters of potato insects: abundance, population growth, and within-plant distribution of the green peach aphid, *Myzus persicae* (Homoptera: Aphididae).*Environmental Entomology* 15: 49-55.

Jeppson.L.R,Keifer.H.H and Baker.W.E 1975. Mites injurious to economic plants. *University of California press* :103-516.

Johnson, W.T. and H.H. Lyon. 1994. Insects that feed on trees and shrubs. *Cornell University Press,* Ithaca, New York. 560 pp.

Joshi,K.C and Agarwal, S.B 1987. Bionomics of the Tent caterpillar, *Malacosoma indica* Wlk.*Bulletin of Entomology*,28(1):1-11.

K. P. Arunkumar, Muralidhar Metta and J. Nagaraju 2006. Molecular phylogeny of silkmoths reveals the origin of domesticated silk moth, *Bombyx mori* from Chinese *Bombyx mandarina* and paternal inheritance of *Antheraea proylei Mitochondrial DNA;* 40 (2): 419–427.

Kailash Chandra Devanshu Gupta, V.P. Uniyal, Abesh K. Sanyal And V. Bhargav 2012.Taxonomic Studies On Lamellicorn Scarabaeids (Coleoptera) Of Simbalbara Wildlife Sanctuary, Sirmour, Himachal Pradesh, India. *Record of. Zoological Survey of India*: 112(1) : 81-91.

Kakar K.L.,Dogra G.S.,Nath A. 1987. Incidence and control of pomegranate fruit borers,*Virachola Isocrates* Fb and *Deudorix epijarbas* Moore. *Indian Journal of Agricultural.Sciences*.57:749-752.

Kambrekar, D.N. and Biradar, A.P. 2015. Field efficacy of insecticides in the management of pomegranate aphid, *aphis punicae* (passerini) *Acta Horticulture.* (ISHS) 1089:153-159.

Kandpal MC and Singh BK 2010. Revised list of Drosophilid species so far described and recorded from India. *Drosophila Information Service*: 93:11-20.

Kanzawa T. 1939. Studies on *Drosophila suzukii* Mats. Kofu. *Review of Applied Entomology* 29: 622.

Katarzyna Michalska , Anna Skoracka, Denise Navia, James W. Amrine. 2009. Behavioral studies on Eriophyid mites: an overview *Experimental and Applied Acarology;* 51(1) :31-59.

Kennedy JS, Day MF, Eastop VF. 1962. A Conspectus of Aphids as Vectors of Plant Viruses. *Commonwealth Institute of Entomology,* London. 114 pp.

Khademi, O. Z. Zamani, Y. Mostofi, S. Kalantari, and A. Ahmadi. 2012. Extending Storability of Persimmon Fruit cv. Karaj by Postharvest Application of Salicylic Acid. *Journal of Agriculture Science and Technology;* 14: 1067-74.

Khan, A. W. 1955. Studies on stocks immune to wooly aphis of apple *Eriosoma lanigerum* (Hausmann). *Punjab Fruit Journal.,* 19: 28-35.

Khan.S.A, Bhatia.S. and Tripathi.N. 2013. Entomological Studies of *Chaetoprocta odata,* An important pest on Walnut trees (Juglans regia L.) in Kashmir valley. *Journal of Academia and Industrial research*:2 (6): 378-381.

Kimura, M. T 1988. Adaptations to temperate climates and evolution of over-wintering strategies in the *Drosophila melanogaster* species group. *Evolution;*42: 1288-1297.

Kosztarab, M. 1963. The armored scale insects of Ohio (Homoptera: Coccoidea: Diaspididae). *Bulletin of the Ohio Biological Survey.*(2) : 2.

Lal, K. B. and Singh, R. N. 1947. Seasonal history and field ecology of the wooly aphid in the Kamaon hills. *Indian Journal Agricultural Sciences,* 17(4): 211-218.

Leconte, J. 1876. Rhincophora of America north of mexico. *Proceedings of the American Philosophical Society;* 15: 224-227.

Loebenstein G, Raccah B. 1980. Control of non-persistently transmitted aphid-borne viruses. *Phytoparasitica* 8: 221-235.

Lowery D.T, Sears MK, Harmer CS. 1990. Control of turnip mosaic virus of rutabaga with applications of oil, whitewash, and insecticides. *Journal of Economic Entomology* 83: 2352-2356.

Lowery DT, Sears M. K. 1986. Effect of exposure to the insecticide azinphosmethyl on reproduction of green peach aphid (Homoptera: Aphididae). *Journal of Economic Entomology;*79: 1534-1538.

Mack TP, Smilowitz Z. 1980. The development of a green peach aphid natural enemy sampling procedure. *Environmental Entomology*: 9: 440-445.

Mackauer M. 1968. Insect parasites of the green peach aphid, *Myzus persicae* Sulz., and their control potential. *Entomophaga;*13: 91-106.

Malik, R. A., Punjabi, A. A. and Bhat, A. A. 1972. Survey study of insect and non-insect pests in Kashmir. *Horticulture,* 3: 29-44.

Mani M., Krishnamoorthy, A and Gupta, P.R. 2013. Egg parasitoids of fruit crop pests. In: Biological control of insect pests using egg parasitoids. *Springer International Publishing Agency.*pp 389-396.

Marco S. 1993. Incidence of nonpersistently transmitted viruses in pepper sprayed with whitewash, oil, and insecticide, alone or combined. *Plant Diseases;*77: 1119-1122.

Marian R. Goldsmith, Toru Shimada and Hiroaki Abe 2005. The genetics and genomics of the silkworm, *Bombyx mori. Annual Review of Entomology* 50: 71

Marshall, G.A.K. 1916. Coleoptera, Rhynchophora: Curculionidae. *Fauna of British India. Taylor and Francis, London.* 367 pp.

Martinat, P.J. 1987. The role of climatic variation and weather in forest insect outbreaks. In: Pedro Barbosa and Jack C. Schultz (eds.) Insect Outbreaks. *Academic Press, Inc., New York.* pp. 241- 268.

Martorell LF. 1945. A survey of the forest insects of Puerto Rico. Part II. *Agricultural. University of Puerto Rico;*29: 399-400.

Masoodi, M. A., and Trali, A. R. 1987. Record of *Chaetoprocta odata* Hewitson (Lycaenida: Lepidoptera) on walnut trees in Kashmir. *Indian Journal of Plant Protection*, 15, 213.

Masoodi, M. A., Vijay, K. K., Bhagat, K. C. and Bhat, O. K. 1989. Biology of Duskey veined walnut aphid (*Callaphis juglandis*) on walnut in Kashmir. *Indian Journal of Plant Protection* 19: 139-141.

Mathew, G. 1997. Management of the Bark Caterpillar *Indarbela quadrinotata* in Forest Plantations of *Paraserianthes falcataria*. *KFRI Research Report* No. 122. Kerala Forest Research Institute, Peechi, India.

Matthews, G.A. 1979. Pesticide application methods. *Longman*, London. 334 pp.

McLeod P. 1991. Influence of temperature on translaminar and systemic toxicities of aphicides for greenpeach aphid (Homoptera: Aphididae) suppression on spinach. *Journal of Economic Entomology*;84: 1558-1561.

McLeod PJ, Steinkraus DC, Correll JC, Morelock TE. 1998. Prevalence of *Erynia neoaphidis* (Entomophthorales: Entomophthoraceae) infections of green peach aphid (Homoptera: Aphididae) on spinach in the Arkansas River Valley. *Environmental Entomology*; 27: 796-800.

Mehta, P.R., and Varma, B.K. 1968. Plant protection. New Delhi, India: Directorate of Extension, Ministry of Food, Agriculture, *Community Development and Cooperation*. 587 pp.

Merrill GB. 1953. Scale insects of Florida. *State Plant Board of Florida Bulletin* 1: 93-94.

Mescheloff, E. and Rosen, D. 1990. Biosystematic studies on the Aphidiidae of Israel (Hymenoptera: Ichneumonoidae). *Israel Journal of Entomology* 24: 35-50.

Michael R. Bush, Michael Klaus, Arthur Antonelli, and Catherine Daniels 1928. Protecting Backyard Apple Trees from Apple Maggot, *Extension Bulletin*.

Miche' Le Roy, Jacques Brodeur, and Conrad Cloutier 1999. Seasonal Abundance of Spider Mites and Their Predators on Red Raspberry in Quebec, Canada *Enviornmental Entomology*; 28(4): 735-747.

Miller, P.W,and Thompson, B.G.1937. Blight and insect pests of walnut. Cooperative extension work in Agriculture and home economics.Oregon State Agriculture college and USDA,USA. *Extension Bulletin* 500.

Milner RJ, Lutton GG.1986. Dependence of *Verticillium lecanii* (Fungi: Hyphomycetes) on high humidities for infection and sporulation using *Myzus persicae* (Homoptera: Aphididae) as host. *Environmental Entomology*. 15: 380-382.

Mir, G.M. and M.A.Wani. 2005. Severity of infestation and damage to walnut plantation by important insect pests in Kashmir. *Indian journal of plant protection* 33(2):188-193.

Misra, C. S. 1920. Index to Indian fruit pests. *Report on the Proceeding of 3rd Entomological Management, Pusa (Bihar)*, 2: 564-595.

Mittler TE, Tsitsipis JA, Kleinjan JE. 1970. Utilization of dehydroascorbic acid and some related compounds by the aphid *Myzus persicae* feeding on an improved diet. *Journal of Insect Physiology*.16: 2315-2326.

Mohinder, S., Gupta, R., and Gupta, P. R. 2007. Biology and control of Indian Gypsy moth, *Lymantria obfuscata* on apple in Himachal Pradesh. *Indian Journal of Plant Protection*, 35(1), 104-105.

Muhammad Sarwar. 2015. Protecting Dried Fruits and Vegetables Against Insect Pests Invasions During Drying and Storage. *American Journal of Marketing Research.* 1(3):142-149.

Müller FP, 1986. The rôle of subspecies in aphids for affairs of applied entomology. *Journal of Applied Entomology*, 101:295-303.

Namba R, Sylvester ES. 1981. Transmission of cauliflower mosaic virus by the green peach, turnip, cabbage, and pea aphids. *Journal of Economic Entomology*.74: 546-551.

Naruse, H. 1978. Defoliation of peach tree caused by the injury of the peach leaf miner, Lyonetia clerkella: Influence of larval density. *Japanese Journal of Applied Entomology and Zoology*. 22: 1-6

Naruse, H., and Hirano, M. 1990. Ecological studies on the peach leaf-miner *Lyonetia clerkella* in the peach field. *Bulletin of Toyama Agricultural Research Centre*, 6: 1-81.

Neuenschwander P, Hagen KS. 1980. Role of the predator *Hemerobius pacificus* in a non-insecticide treated artichoke field. *Environmental Entomology*.9: 492-495.

Nowierski, R. M., A. P. Gutierrez, and J. S. Yaninek. 1983. Estimation of thermal thresholds and age-specific life table parameters for the walnut aphid (Homoptera: Aphididae) under field conditions. *Environmental Entomology*. 12:680–686.

Oatman E. R., Legner E. F. 1961. Bionomics of the apple aphid, Aphis pomi, on young non-bearing apple trees. *Journal of Economic Entomology*. 54: 1034-1037.

O'Brien, C.W., M. Haseeb and M.C. Thomas. 2006. *Myllocerus undecimpustulatus undatus* Marshall (Coleoptera: Curculionidae), a recently discovered pest weevil from the Indian subcontinent. *Florida Department of Agriculture and Consumer Services.* Division of Plant Industry. Entomology Circular No. 412: 1-4.

Palmer MA. 1952. Aphids of the Rocky Mountain Region. *Thomas Say Foundation*, 5. 452 pp.

Palumbo JC, Kerns DL. 1994. Effects of imidacloprid as a soil treatment on colonization of green peach aphid and marketability of lettuce. Southwestern Entomologist 19: 339-346.

Pandey. A. K. and Mali P. 2016. Insect pest management of fruit crops : Biotech Books. 689 pp.

Panwar, V. P. S. 2002. Agricultural insect pests of crop and their control, pp. 164-175.

Pawar, A. D., Tuhan, N. C., Balsubramanian, S. and Parry, M. 1981. Distribution, damage and biology of codling moth, *Cydia pomonella* (L). *Indian Journal of Plant Protection* 10: 111-114.

Petitt FL, Smilowitz Z. 1982. Green peach aphid feeding damage to potato in various plant growth stages. *Journal of Economic Entomology*,75: 431-435.

Phelan P, Montgomery ME, Nault LR. 1976. Orientation and locomotion of apterous aphids dislodged from their hosts by alarm pheromone. *Annals of the Entomological Society of America*, 69: 1153-1156.

Powell D. M, Mondor WT. 1976. Area control of the green peach aphid on peach and the reduction of potato leaf roll virus. *American Potato Journal* 53: 123-139.

Powell D. M. 1980. Control of the green peach aphid on potatoes with soil systemic insecticides: pre-plant broadcast and planting time furrow applications.*Journal of Economic Entomology*;73: 839-843.

Pritts, M. P., Handley, D. 1989. Natural Resource, Agricultural and Engineering Service (NRAES), *Bramble Production Guide*.

Pritts, M. P. 1999. Raspberries in winter: An Alternative Growing System for Northern States. *New York Fruit Quarterly*. 6(4): 2-4,

Pritts, M. P., W. Langhans, T. K. Whitlow, M. J. Kelly and A. Roberts. 1999. Winter Raspberry Production in Greenhouses. *Horticulture Technology*, 9(1): 13-15,

Pritts, M. P. 2000. Winter Raspberry Production in a Greenhouse. *Fruit Notes*, 65: 20-21.

Pruthi, H. S. 1938. The distribution, status and biology of codling moth (*Cydia pomonella* L.) in Baluchistan with notes on some other insects affecting apple. *Indian Journal of Agricultural Sciences*, 9: 499-54.

Quayle, H. J. 1938. Insects of citrus and other subtropical fruits. *Comstock Publishing Co. Ithaca New York.*

Radloff, Sarah E. 2010. Population structure and classification of *Apis cerana.,Apidologie* 41.6 : 589-601.

Rahman, K. A. and Ansari, R. A. 1941. Scale insects of the Punjab and northwest frontier province usually mistaken for San Jose (with description of the new species). *Indian Journal of Agricultural Sciences*. 14(4): 308-310.

Rahman, K. A. Asa, N. and Kalia, 1940. Volant animals which act as carriers of San Jose scale. *Current Science*. 9(5): 235.

Rahman, Khan A. 1941. Occurrence of the gypsy moth, *Lymantria obfuscate* Walk. In Shimla hills. *Indian Journal of Entomology*. 31(2): 338.

Rahman, Khan A. and Abdul Wahid Khan 1941. Biology and control of wooly aphid *Eriosoma lanigerum* Hausmann (Aphididae: Rhynchota) in the Punjab. *Indian Journal of Agricultural Sciences* 11(2): 265.

Ramamurthy, V.V. and S. Ghai. 1988. A study on the genus *Myllocerus* (Coleoptera: Curculionidae). *Oriental Insects*, 22: 377-500.

Rexrode, C.O. 1982. Bionomics of the peach bark beetle *Phloeotribus liminaris* (Coleoptera: Scolytidae) in black cherry. *Journal of the Georgia Entomological Society* 17:388-398.

Rice, M.E. 1995. Branch girdling by *Oncideres cingulata* (Coleoptera: Cerambycidae) and relative host quality of persimmon, hickory, and elm. *Annals of the Entomological Society of America*.88:451-455.

Ring, D.R., L.J. Grauke, J.A. Payne, and J.W. Snow. 1991. Tree species used as hosts by pecan weevil (Coleoptera: Curculionidae). *Journal of Economic Entomology*, 84:1782-1789.

Ruchie Gupta and J. S. Tara. 2013. Biological Studies of *Lymantria Obfuscata* Walker (Lepidoptera: Lymantridae) On Apple Plantations (*Malus Domestica* Borkh.) In Jammu Region of J&K, India.*Munis Entomology and Zoology*, 8(2): 749

S. K. Javorekac, K. E. Mackenziec, S. P. Vander Kloetbc 2002. Comparative pollination effectiveness among bees (Hymenoptera: Apoidea) on low bush blueberry. *Annals of the Entomological Society of America*.95 (3): 345–351.

Sasidharan K.R. and Varma. R.V. 2008. Seasonal population variations of the bark eating caterpillar (*Indarbela quadrinotata*) in casuarina plantations of Tamil nadu. *Tropical ecology*,49(1): 79-83,

Sasidharan, K.R. 2004. Studies on the Insect Pests of *Casuarina equisetifolia* L. in Tamil Nadu and their Management. Ph.D. Thesis, *Forest Research Institute, Deemed University, Dehradun, India.*

Schaefer, C.W. and A.R. Panizzi. 2000. Heteroptera of Economic Importance. CRC Press, *Boca Raton.* 828 p.

Schultz, P. B. 1984. Natural enemies of oak Lecanium (Homoptera: Coccidae) in eastern Virginia. *Environmental Entomology*, 13: 1515–1518.

Sedlacek, J. D., K. V. Yeargan, and P. H. Freytag. 1986. Laboratory life table studies of the blackfaced leafhopper (Homoptera: Cicadellidae) on Johnson grass and corn. *Enviornmental. Entomology*, 15:1119-1123.

Shah.R,Fazal.H.and Poswal.M.A. 2012. Ontogeny and integrated management of *Alcidodes porrectirostris* Marsha (Coleoptera: Curculionidae) infesting walnut fruits in Manoor Valley (Kaghan), *KPK, Pakistan Journal of Entomological Research.* 36(1):1-7.

Shanks, C. H., Jr., A. L. Antonelli, and B. D. Congdon. 1992. Effect of pesticides on twospotted spider mite (Acari: Tetranychidae) populations on red raspberries in Western Washington. *Agriculture Ecosystem and Environment*, 38: 159 -165.

Shean B, Cranshaw WS. 1991. Differential susceptibilities of green peach aphid (Homoptera: Aphididae) and two endoparasitoids (Hymenoptera: Encyrtidae and Braconidae) to pesticides. *Journal of Economic Entomology*,84: 844-850.

Sheikh, A. G. 1975. Effect of repeated defoliation caused by *Lymantria obfuscata* on apple trees in Kashmir. *Himalayan Horticulture*, 5 (3/4): 9-29.

Sheikh, A. G. 1985. Insect pests of temperate fruits and their management. *National Workshop cum Seminar on Temperate Fruits*. SKUAST pp. 95-98.

Sheldeshova, G. G. 1967. Ecological factors determining distribution of codling moth, *Laspeyria pomonella* L (Lepidoptera: Torticidae) in northern and southern hemispheres. *Entomological Review*, 46: 349-361.

Simonet, D. E. and R. L. Pienkowski. 1980. Temperature effect on development and morhometrics of the potato leafhopper. *Environmental Entomology*. 9:798–800.

Singh, C. 1964. Temperate fruit pests. In: Entomology in India: *Entomological Society of India*, New Delhi. 213-276.

Singh, R., Kumar, S., Chakrabarthy and Kumar, A. 2007. Resurgence of Indian gypsy moth, *Lymantria obfuscata* Walker (Lepidoptera: Lymantriidae) on ban oak (*Quercus leucotrichophora*) forests in Rajgarh forest division, Himachal Pradesh. *Indian Journal of Forestry*, 30 (1): 83-85.

Singh, S. S., Gupta, N.N.and Rai, M.K. 2010. Pest management in temperate fruits: Opportunities and Challenges. *Progressive Horticulture*,42(1):76-83.

Srivastava K.P. 1996. Text book of Applied Entomology: *Kalyani Publishers*. pp. 140-162.

Srivastava, A. S. and Masoodi, M. A. 1985. Influence of host plants on the development and survival of *Lymantria obfuscata*. F.A.O. *Plant Protection Bulletin*, 33 (2): 67-69.

Stamp, N. 2003. Out of the quagmire of plant defense hypotheses. *The Quarterly Review of Biology*,78: 23- 55.

Stewart JK, Aharoni Y, Hartsell PL, Young DK. 1980. Acetaldehyde fumigation at reduced pressures to control the green peach aphid on wrapped and packed head lettuce. *Journal of Economic Entomology*,73: 149-152.

Stoetzel MB, Miller GL, O'Brien PJ, Graves JB. 1996. Aphids (Homoptera: Aphididae) colonizing cotton in the United States. *Florida Entomologist*, 79: 193-205.

Storey W.B., Enderund J.E., Saleeb W.S. and Nauer E.M. 1977. The Fig (*Ficus carica* Linnaeus). Its Biology, History, Culture and Utilisation. *Jurupa Mountains Cultural Centre, California USA.*

Sunil B. Avhad and Chandrashekar J. Hiware 2013. Mulberry Defoliators: Distribution and Occurrence from Aurangabad (M.S.), *Journal of Entomology and Zoological studies*.1 (4):1-6.

Swirski, E. and Amitai, S. 1999. Annotated list of aphids (Aphidoidea) in Israel. *Israel Journal of Entomology* **33**: 1-120.

Tamaki G, Annis B, Fox L, Gupta RK, Meszleny A. 1982. Comparison of yellow holocyclic and green anholocyclic strains of *Myzus persicae* (Sulzer): low temperature adaptability. *Environmental Entomology*,11: 231-233.

Tamaki G, Annis B, Weiss M. 1981. Response of natural enemies to the green peach aphid in different plant cultures. *Environmental Entomology*, 10: 375-378.

Tamaki G, Fox L. 1982. Weed species hosting viruliferous green peach aphids, vector of beet western yellows virus. *Environmental Entomology*,11: 115-117.

Tamaki G, Halfhill JE. 1968. Bands on peach trees as shelters for predators of the green peach aphid. *Journal of Economic Entomology*,61: 707-711.

Tamaki G. 1975. Weeds in orchards as important alternate sources of green peach aphids in late spring. *Environmental Entomology*,4: 958-960.

Tan, K.H. and Serit, M. 1994. Adult population dynamics of *Bactrocera dorsalis* (Diptera: Tephritidae) in relation to host phenology and weather in two villages of Penang Island, Malaysia. *Environmental Entomology*,23(2): 267-275.

Tingey WM, Laubengayer JE. 1981. Defense against the green peach aphid and potato leafhopper by glandular trichomes of *Solanum berthaultii*. *Journal of Economic Entomology*,74: 721-725.

Umesh KC, Valencia J, Hurley C, Gubler WD, Falk BW. 1995. Stylet oil provides limited control of aphid-transmitted viruses in melons. *California Agriculture*,49: 22-24.

V. V. Ramamurthy, V.S. Singh, G.P. Gupta and A. V. N. Paul 2005. *Gleanings in Entomology*. Division of Entomology, Indian Agricultural Research Institute, New Delhi 110012. pp. 317.

Van Emden HF, Eastop VF, Hughes RD, Way MJ. 1969. The ecology of *Myzus persicae*. *Annual Review of Entomology*,14: 197-270.

Vasantharaj David. 2001. *Elements of Economic Entomology*. Popular Book Depot, Chennai. pp.562.

Verghese, A. and Jayanthi, P.D.K. 2001. Integrated pest management in major fruit crops.In: P.Parvatha reddy, A. Verghese and N.K. Kumar (eds.), *Integrated pest management in Horticultural Ecosystem*, Capital Publishing Company, New Delhi-110002. pp.12-15.

Verma. T.D.1985. Incidence and chemical control of bark- eating caterpillar, *Indarbela quadrinotata* Walker on plum trees. *Indian Journal of Agricultural Sciences*. 55(2): 131-132.

Vigneshwara, V. 2012. Walnuts: Need to expand cultivation. *Market Survey*.pp. 13-15.

Wadallah KT, Ammar ED, Tawafik MFS, Rashad A, 1979. Life history of the white mealybug *Ferrisia virgata* (Ckll.) (Homoptera: Pseudococcidae). *Zeitschrift für Deutsche Entomologen*, 26:101-110.

Wadhi, S. R. and Sethi, G. R. 1975. Eradication of codling moth-a suggestion. *Journal of Nuclear Agriculture and Biology*. 4: 14-19.

Waggoner, M., Ohr, H., Adams, J., and Gonzalez, D. 2000. Methyl iodide: an alternative to methyl bromide for insectary fumigation. *Journal of Applied Entomology* 124:113-117.

Waliullah, M. I. S. 1992. Nematodes associated with vegetables crops in the Kashmir Valley, India. *Nematologia Mediterranea*, 20 (1):47-48.

Webb, R. E., Peiffer, R., Fuester, R. W., Thorpe, K. W., Calabrese, L., and McLaughlin, J. M. 1998. An evaluation of the residual activity of traditional, safe, and biological insecticides against the gypsy moth. *Journal of Arboriculture*, 24(5), 286-293.

White AJ, Wratten SD, Berry NA, Weigmann U. 1995. Habitat manipulation to enhance biological control of *Brassica* pests by hover flies (Diptera: Syrphidae). *Journal of Economic Entomology*,88: 1171-1176.

Williams, D. J. 1996. A brief account of the hibiscus mealy bug *Maconellicoccus hirsutus*, a pest of agriculture and horticulture, with descriptions of two related species from southern Asia. *Bulletin of Entomological Research*, 86 :617–628.

Williams, D. and A. Liebhold. 1995. Forest defoliators and climatic change: Potential changes in spatial distribution of outbreaks of western spruce budworm (Lepidoptera: Tortricidae) and gypsy moth (Lepidoptera: Lymantriidae). *Environmental Entomology*, 24: 1-9.

Winstanley, J.K. The fig tree borer, *Phryneta spinator* F. (Cerambycidae: Coleoptera). Pests and Diseases of South African Forests and Timber: *Pamphlet* 273.

Wolfenbarger, D.O. 1972. Effects of temperatures on mortality of green peach aphids on potatoes treated with ethyl-methyl parathion. *Journal of Economic Entomology* 65: 881-882.

Wyman JA, Toscano NC, Kido K, Johnson H, Mayberry KS. 1979. Effects of mulching on the spread of aphid-transmitted watermelon mosaic virus to summer squash. *Journal of Economic Entomology*, 72: 139-143.

Yaku A, Walter GH, Najar- Rodriguez AJ. 2007. Thrips red flower colour and the host relationships of a polyphagous anthophilic thrips. *Ecological Entomology*, 32: 527-535.

Zaki, F. A, and Chan, S. A. 2001. Population build-up pattern of ERM *Panonychus ulmi* in managed and unmanaged apple orchards in Kashmir. *SKUAST Journal of Research* 3: 9-13.

Zaki, F.A. 1999. Incidence and biology of codling moth, *Cydia pomonella* L., in Ladakh (Jammu and Kashmir). *Applied Biological Research*.1: 75-78.

Zalom, F. G. 1981. Effects of aluminum mulch on fecundity of apterous *Myzus persicae* on head lettuce in a field planting. *Entomologica Experimentalis et Appliciata*, 30: 227-230.

Select Uniform Resource Locator (URL)

http://www.extento.hawaii.edu/kbase/crop/Type/c_viridi.htm

http://afghanag.ucdavis.edu/a_horticulture/fruits-trees/figs 1/manfruitnswfiggrowingext.pdf

http://agropedia.iitk.ac.in/content/pomegranate-mealy-bug

http://www.extento.hawaii.edu/kbase/crop/type/bloss_midgei.htm

http://www.ipm.ucdavis.edu/PMG/selectnewpest.kiwifruit.html

http://www.yourarticlelibrary.com/plant-diseases/insects-and-diseases-that-occurs-in-loquat-plants-and-measures-to-control-it/24694/

https://en.wikipedia.org/wiki/*Lasioderma_serricorne*

http//Alert List/insects/*drosophila suzukii.*

http://agricoop.nic.in/dacdivision/machinery1/chap4.pdf

http://agridr.in/tnauEAgri/eagri50/ENTO331/lecture09/castor/005.html

http://agritech.tnau.ac.in/crop_protection/cocoa/cocoa_6.html

http://agritech.tnau.ac.in/crop_protection/pests per cent 20of per cent 20pome.html

http://agropedia.iitk.ac.in/content/pomegranate-bark-eating-caterpillar

http://agropedia.iitk.ac.in/content/pomegranate-bark-eating-caterpillar.

http://agropedia.iitk.ac.in/content/pomegranate-fruit-borer.

http://axp.ipm.ucdavis.edu/PMG/select new pest.strawberry.html.

http://cherries.msu.edu/uploads/files/PDFs/Insects/PlumRustMite.pdf

http://elmostreport.blogspot.in/2011/01/round-headed-apple-tree-borer-saperda.html.

http://entnemdept.ufl.edu/creatures/FRUIT/strawberry_leafroller.htm

http://entnemdept.ufl.edu/creatures/orn/scales/green_scale.htm

http://ento.psu.edu/extension/factsheets/black-cutworm.

http://entomology.ifas.ufl.edu/walker/ufbir/chap15.htm.

http://extension.missouri.edu/p/G7190

http://extension.missouri.edu/p/G7190. Insect Borers of Fruit Trees

http://homeguides.sfgate.com/nematodes-apple-trees-23021.html

http://homeguides.sfgate.com/raspberry-leaves-aphids-22667.html

http://homeguides.sfgate.com/use-rid-aphids-pomegranate-tree-61434.html

http://insect.pnwhandbooks.org/tree-fruit/plum-and-prune/plum-and-prune-rust mite

http://ipm.ncsu.edu/AG136/leafhop1.html.potato leaf hopper.

http://jenny.tfrec.wsu.edu/opm/displaySpecies.php?pn=320

http://jenny.tfrec.wsu.edu/opm/displaySpecies.php?pn=420.Black cherry aphid.

http://jenny.tfrec.wsu.edu/opm/displayspecies.php?pn=640

http://kiengiangbiospherereserve.com.vn/doc/12.Control_bark_eating-caterpillar.pdf

http://lee.ifas.ufl.edu/Hort/UsefulLawnandGardenResources/FigFruitFlyupdate2008.pdf

http://msue.anr.msu.edu/topic/chestnuts/pest_management/insects_and_pests

http://niphm.gov.in/IPMPackages/Pomegranate.pdf

http://poorjavad.iut.ac.ir/biology-and-seasonal-fluctuations-pomegranate-aphid-aphis-punicae

http://www.agf.gov.bc.ca/cropprot/swd.htm.

http://www.agri.huji.ac.il/mepests/pest/Aphis_punicae.

http://www.agri.huji.ac.il/mepests/pest/*Batocera rufomaculata.*

http://www.agri.huji.ac.il/mepests/pest/Ferrisia_virgata.

http://www.agric.nsw.gov.au/Hort/ascu/insects/ashwf.htm

http://www.banglajol.info/index.php/BJAR/article/view/27666

http://www.biotecharticles.com/Pest-Management-of-Grapevine-Stem-Borer-andGirdler

http://www.cdfa.ca.gov/phpps/ipc/biocontrol/83ash white fly.htm.

http://www.doacs.state.fl.us/pi/enpp/ento/weevil-pestalert.html.

http://www.freshfromflorida.com/content/download/23897/486234/siphoninus-phillyreae.pdf

http://www.fruit.cornell.edu/Berries/hightunnels.html.

http://www.fruitipedia.com/Persimmon.htm.

http://www.gardenguides.com/101903-rid-raspberry-plants-mites.html

http://www.inaturalist.org/taxa/american_rose_chafer.

http://www.ipm.ucdavis.edu/PDF/PMG/pmgwalnut.pdf.

http://www.ipm.ucdavis.edu/PMG/PESTNOTES/pn74158.html

http://www.ipm.ucdavis.edu/PMG/PESTNOTES/pn7489.html

http://www.ipm.ucdavis.edu/PMG/r105301811.html

http://www.ipm.ucdavis.edu/PMG/r4200111.html

http://www.ipm.ucdavis.edu/PMG/r621300211.html

http://www.ipm.ucdavis.edu/PMG/r881300511.html

http://www.na.fs.fed.us/spfo/pubs/fidls/beechbark/fidl-beech.htm.

http://www.na.fs.fed.us/spfo/pubs/fidls/gypsymoth/gypsy.htm.

http://www.virginiafruit.ento.vt.edu/GFB.html. Grape flea beetle.

http://www.virginiafruit.ento.vt.edu/StrwWeevil.html

http://www1.agric.gov.ab.ca/$department/deptdocs.nsf/all/prm2545

http://www-museum.unl.edu/research/entomology/Guide/index4.htm).

http://wwwmuseum.unl.edu/research/entomologyGuide/Scarabaeoideapages/Scarabaeoidea-

https://en.wikipedia.org/wiki/Castanea_sativa.

https://en.wikipedia.org/wiki/Chestnut

https://en.wikipedia.org/wiki/Chestnut.

https://en.wikipedia.org/wiki/Honey_bee

https://extension.tennessee.edu/publications/documents/SP503-I.pdf

https://www.plantvillage.org>topics>Quince diseases and pests.

https://www.rhs.org.uk/advice/profile?PID=201

www.gardeningknowhow.com/plant-problems/pests/insects/controlling-tortrix-moths.

Glossary

Abdomen: The abdomen is one of the three main body segments of insects in the posterior part with segmentation. The abdomen contains the heart, reproductive organs, midgut and other digestive organs.

Abscise: To separate by abscission as a leaf from a stem.

Abscission: The act of cutting off or sudden termination.

Adult: Adult is the name given to the imago stage of an insect.it is the stage when an insect is sexually mature and ready to reproduce normally.

Aesthetic: A set of ideas or opinion about beauty or art.

Aestivation: A period of summer dormancy.

Alate: Having or possessing wings.

Androecia: (singular = Androconium). In male butterflies, specialized wing scales (often called scent scales) possessing special glands, which produce a chemical attractive to females.

Anholocyclic: (obligatory parthenogenesis). Pertaining to an insect that does not undergo sexual reproduction during its life cycle.

Annulate: Formed in ring like segments or with ring like markings.

Antennae: A pair of sensory organs in insects. The long feelers situated on the head and close to the eyes. They are however not tactile but used for detecting airborne scents and currents.

Anterior: In front of or after the aforementioned structure.

Anthesis: The flowering period of a plant from the opening of flower bud.

Antioxidant: A substance that inhibits oxidation.

Aphid: Insects in the Superfamily Aphidoidea within the Suborder Sternorrhyncha. They are often considered pests on plants.

Apices: (plural of apex). Meaning the tip.

Apterous: Without wings.

Arachnid: The Class of Arthropods that includes spiders, mites, ticks, scorpions, pseudo-scorpions and harvestmen.

Arrhenotoky: The type of parthenogenesis where only males are produced.

Atrophy: Atrophy is the partial or complete wasting away of a part of the body. It is the physiological process of reabsorption and breakdown of tissues

Axil: The point at which thoracic muscles attach to the wing of an insect.

Bait: A food or lure used in trap.

Beetle: An insect distinguished by having forewings that are modified into hard wing cases. Beetles are regarded as the most species rich of all the Orders of insects.

Berry: A berry is a fleshy fruit without a stone produced from a single flower containing one ovary.

Bio fix: A biological event or indicator of a developmental event, usually in the life of an insect pest, which initiates the beginning of growing degree-day calculations.

Biparental: Pertaining to or having traits or characteristics that stem from both parents.

Black cap stage: Overwintering stage of San Jose scale in which the waxy covering turns from white to black.

Blemish: A small mark that makes the appearance of something less attractive.

Blister: Gall mite sometimes severely injures the leaves while feeding forms blisters.

Blossom: A flower or a mass of flowers, especially on a tree or bush.

Blotch: Any plant diseases caused by fungi or bacteria and resulting in brown or black dead areas on leaves or fruit.

Bramble: A prickly shrub of the genus Rubus of the rose family, including the blackberry and the raspberry.

Brood: A number of young ones produced or hatched at one time.

Bud burst: The emergence of new leaves on a plant at the beginning of each growing season.

Bud scales: A bud scale is a kind of specialized leaf or bract that protects and surrounds a dormant plant bud before the bud expands.

Bug: The Hemiptera/true bugs are an order of insects including the aphids.

Bump: Raised area on the tree.

Bunch: A connected group or cluster.

Bunchy top: Bunchy top is a viral disease caused by a single-stranded DNA virus called the bunchy top virus.

Burlap sticks: Sticks used to keep pests off plants with barriers on tree trunk.

Cache: A hiding place.

Calyx: The outer part of a flower formed by sepals.

Cambium: A tissue layer that provides partially undifferentiated cells for plant growth.

Camouflage: The ability for a organism to blend in with its environment.

Cannibalistic: When insect predates upon another insect of its own species.

Canopy: Upper most branches of the tree.

Caterpillar: The second stage in the life cycle of butterflies and moths.

Chafer: A family of beetles that has distinctively shaped antennae and is often metallic in colouration.

Chelicerae: Jointed appendages possessed by some arthropods and used for feeding.

Chlorosis: Loss of normal green coloration of leaves of plants.

Clump: A small group of trees or plants growing closely together.

Cluster: A bunch.

cm: Centimeter.

Cocoon: The protective covering around the pupae or chrysalis of some insects.

Coleoptera: Sheath winged: an order with the anterior wings coriaceous, used as a cover only, meeting in a straight line dorsally, mouth mandibulate, prothorax free, transformations complete; the beetles; the term has also been applied to the two elytra together.

Concave: Curved like the inner surface of a sphere.

Congregate: To come together or assemble in large number.

Conspicuous: Readily visible or easily seen.

Convex: Curved or rounded outward.

Crawler: Early form of insect larvae.

Crescent: Having the shape of a crescent.

Crevice: A narrow opening resulting from a split or crack.

Crinkling: Form into small surface creases or wrinkles.

Crotch: The region formed by the junction of two parts such as two legs or branches.

Crown: Rest on or form the top of.

Cruciform: Having the shape of a cross.

Cultivar: A plant variety that has been produced in cultivation by selective breeding.

Curculio: A beetle of the weevil family especially one which is a pest of fruit trees.

Cylindrical: Having the form or properties of a cylinder.

Dead wood: A branch or part of tree, which is dead.

Debris: The pieces that are left after something has been destroyed.

Deciduous: Falling off at maturity or tending to fall off.

Defoliate: Remove leaves from.

Degree-day: Degree days are used to predict insect life cycles.

Desiccation: Ability of an organism to withstand or endure extreme dryness.

Deutogynes: Overwintering females.

Deutonymph: A second larval form occurring in the development of most mites.

Development threshold: Threshold of development. A given temperature below which no development of an organism may take place.

Diametrical: Of or along a diameter.

Diapause: The arrested development of an organism. Diapause is often the result of environmental conditions.

Diatomaceous earth: It is a naturally occurring soft siliceous sedimentary rock that is easily crumbled into a fine white to off white powder.

Dieback: A condition in which a tree or shrub begins to die.

Diptera: An ordinal term applied to insects having only one pair of wings(anterior); thorax agglutinate; mouth haustellate; transformations complete.

Dorsum: The dorsal part of an organism.

Druplets: A term druplet is used for the aggregate fruit of raspberries and blackberries.

Dusk: Dusk is the darkest part of the evening twilight.

EC: Emulsifiable concentrate.

Eclosion: the emergence of an insect from a pupa or egg.

Egg sac: The pouche in which insect deposit their eggs. It is also called egg case.

Egg scars: Egg laying by the overwintering adults misshape the fruit resulting in egg scars.

Elliptical: Shaped like a flattened circle.

Elongate: To be stretched out.

Elytra: Each of the two wing cases of a beetle.

Embryonic: In the state of an embryo.

Epidermal: The protective outer layer of the skin.

Epithelium: Thin tissue forming the outer layer of a body's surface.

Erineum: An abnormal felty growth of hairs from the leaf epidermis of plants caused by various mites.

Eriophyid: These are the microscopic mites having only two pair of legs. These include bud, blister, gall and rust mites.

Exacerbate: To increase the severity.

Excavate: To uncover something by digging.

Excreta: Waste matter discharged from the body.

Exude: Discharge of insect body.

Faeces: Waste matter remaining after food has been digested.

Family: In zoological classification, a level of the taxonomic hierarchy below the order and above the genus. All zoological family names end in the suffix—idea.

Flagging: Mark an item for attention or treatment in a simple way.

Fleck: Mark with small patches of colour or particles of something.

Florescence: The process of flowering.

Foliage: A cluster of leaves.

Forewings: Either of the two front wings of a four winged insect.

Forking: Dig or move anything with a fork.

Frass: The droppings of insect larvae.

Fringe: The outer or less important part of an area. In order *Thysanoptera thysanos* means fringe and ptera means wings.

Froth: A mass of small bubbles in liquid caused by salivating. The insects producing froth are known as spittlebugs.

Fruit drop: The shedding of unripe fruit from a tree.

Fruitlet: An immature or small fruit.

G: Granules.

Gall mite: A minute mite, which is parasitic on plants, typically living inside buds and causing them to form hard galls.

Galls: A growth on a plant in response to the action of an insect.

Generation: The population born and living at about the same time.

Girdle: A silken thread used to support the pupae of a butterfly or moth.

Glossy: Shiny and smooth.

Gnathosoma: It is the part of the body of mites and ticks comprising the mouth and feeding parts. These are the hypostome, the chelicerae and pedipalp.

Gnaw: Bite at or nibble something persistently.

Gravid: (Pregnant). Carrying eggs or young ones.

Green tip stage: It describes the stage following silver tip when the buds are broken at the tip and about 1/16 inch of green tissue is visible.

Gregarious: Living in flocks or loosely organized communities.

Grub: Thick bodied sluggish larva of several insects as of scarab beetle.

Gynoecia: The female part of a flower, consisting of one or more carpels.

Hard shells: Outer covering of some insects to protect them.

Heartwood: The older harder non living central wood of trees that is darker denser, less permeable and more durable than the surrounding sapwood.

Hemiptera: Half winged; an ordinal term applied to insects in which the mouth parts consist of four lancets enclosed in a jointed beak or rostrum; metamorphosis incomplete; the forewings may be of uniform texture throughout or they may be thickened at the base and membranous at the tip.

Hibernaculum: It is a place where insect hibernate to survive the winter.

Hibernation: A period of winter dormancy.

Hind leg: One of two rear limbs or the posterior limb on an insect, especially the quadrupeds.

Hoeing: Thin out or dig up.

Honeydew: It is a sugar-rich sticky liquid, secreted by aphids and some scale insects as they feed on plant sap.

Hopper burn: Browning or shriveling of foliage associated with the feeding of leafhopper.

Hover: To stay in one place in air.

Hull/Husk: The dry outer covering of a fruit, seed, or nut.

Hymenoptera: It is the third largest order of insects that includes bees, wasps, ants, and sawflies. They have four transparent wings and the females typically have the sting.

Hyper parasites: An organism that is parasitic on other parasite.

Imago: The adult insect.

In situ: In the original place.

Incision: A cut made on flesh.

Indentation: It is a notch, cut, or deep recess on the edge or surface of something.

Innocuous: Not harmful or offensive.

Insect pests: Insect pests are those insects that feed on, compete for food with, or transmit diseases to humans and livestock.

Instar: An immature arthropod between moults.

Internode: An interval or part between two nodes.

Iridescent: Shining with many different colors when seen from different angles.

Juvenile: Physiologically immature or undeveloped.

Kernel: The small, somewhat soft part inside a seed or nut.

Lacerate: To cut or tear deeply or roughly.

Larva: The juvenile form of an insect.

Lateral: Relating to the side.

Latex: White fluid produced by certain plants that is used for making rubber.

Leaf lamina: A leaf consists of a broad, expanded blade (the lamina), attached to the plant stem by a stalk like petiole.

Leaf petiole: The petiole is the stalk that attaches the leaf blade to the stem.

Leaf wilt: The drying out, drooping, and withering of the leaves of a plant due to inadequate water supply, excessive transpiration, or vascular disease.

Lepidoptera: Scale winged; an order of insects with spirally coiled haustellate mouth structures, head free, thorax agglutinate, transformations complete, four scale covered wings.

Lesion: The abnormal change in the tissue.

Limb: One of the projecting paired appendages of an animal body used especially for movement and grasping but sometimes modified into sensory or sexual organs.

Litter: The offspring at one birth of a multiparous animal.

Lure trap: A decoy used in catching insect's especially artificial bait.

Maggot: A soft bodied legless grub that is the larvae of a dipterous insect.

Malformed: (Deformed) not having the normal or expected shape especially because of a problem in the way something has developed or grown.

Mandible: A part of an insect's mouth that looks like a jaw and is often used for biting things.

Mate: Either member of a breeding pair of animals.

Mealy bug: Any of a family of scale insects that have a white cottony or waxy covering and are destructive pests especially of fruit trees.

Mealy: Flecked with another colour.

Mesophyll: The parenchyma between the epidermal layers of a foliage leaf.

Methyl eugenol: It is a phenylpropene, a type of phenylpropanoid compound, the methyl ether of eugenol. It is found in various essential oils.

Midge: Midges are a group of insects that include many kinds of small flies.

Midrib: A larger strengthened vein along the midline of a leaf.

Mimic: Imitative of something.

Mite: A group of small arachnids that have a very diverse habit ranging from parasitism to free-living organisms.

mm: Millimeter.

Moldy: Covered with a fungal growth, which causes decay, due to age or damp conditions.

Monophagous: Feeding on a single kind of plant or animal.

Mosaic virus: It is a plant virus that causes the leaves to have a speckled appearance.

Moth: Butterflies and Moths come under the Order Lepidoptera are probably the most widely studied order of invertebrates. The Order comprises some 160,000 species.

Mottle: It is a pattern of irregular marks, spots, streaks, blotchs or patches of different shades or colours.

Moult: A loss of feathers, hair, or skin, especially as a regular feature of an animal's life cycle.

Mow: To cut down.

Mulch: Mulch is a layer of material applied to the surface of an area of soil. Its purpose is to conserve moisture, to improve the fertility and health of the soil, to reduce weed growth.

Mummify: Dry up and so preserve it.

Mycoplasma: It is a genus of bacteria that lack a cell wall around their cell membrane.

Native: Plant or animal of indigenous origin or growth.

Necrosis: The death of most or all of the cells in an organ or tissue due to disease, injury, or failure of the blood supply.

Nectar: Nectar is sugar-rich liquid produced by plants in glands called nectaries, either within the flowers with which it attracts.

Nectarine: A peach of a variety with smooth red and yellow skin and rich, firm flesh.

Neonate: New born.

Nibble: Take small bites out of.

Niche: The place or function of a given organism within its ecosystem.

Nocturnal: Active at night.

Non-host: A plant that is not parasitized by a particular organism.

Notching: Slit in an object, surface or edge.

Nutlets: Small nut like fruit or seed.

Nymph: The juvenile form of an insect.

Obtect: Insect pupae covered in a hard case with the legs and wings attached immovably against the body.

Offspring: Young born of living organism, produced either by a single organism or in the case of sexual reproduction, two organisms.

Oligophagous: Feeding on a limited number of foods.

Omnivorous: Feeding on a variety of food of both plant and animal origin.

Ooze: The sluggish flow of a fluid.

Opisthosoma: The opisthosoma is the posterior part of the body in some arthropods, behind the propodosoma.

Oval: Having a rounded and slightly elongate outline or shape like that of an egg.

Overlapping: To lap over.

Over-winter: Live through the winter.

Oviparous: Producing young by means of eggs, which are hatched after they have been laid by the parents.

Oviposition: The act of laying eggs.

Ovisac: A sac or capsule containing an ovum or ova.

Ovoid: Shaped like an egg.

Ovoviviparous: Producing young by means of eggs, which are hatched within the body of the parent.

Palmate: A type of compound leaf with small leaves that all grow from the same point at the end of the stem.

Parasite: An organism, which lives in or on another organism and benefits by deriving nutrients and protection from it.

Parasitoid: An insect whose larvae live as parasite, which eventually kill their hosts.

Parenchyma: It comprises the functional parts of an organ and in plants it is the ground tissue of non-woody structures.

Parthenogenesis: Reproduction from an ovum without fertilization. It is the mode of asexual reproduction in which females without the genetic contribution of a male produce offspring's.

Pedicel: Small stalks bearing an individual flower in an inflorescence.

Perforate: Make a hole by boring or punching.

Peristalsis: It is a series of wave like muscle contractions that moves food to different processing stations in the digestive tract.

Petal fall: The last stage of flower development in fruit trees.

Petal: Each of the segments of the corolla of a flower, which are modified leaves and are typically coloured.

Petiole: The stalk that joins a leaf to a stem.

Phenology: It is the study of periodic plant and animal life cycle events and how these are influenced by seasonal and interannual variations in climate, as well as habitat factors.

Pheromone trap: It is a type of insect trap that uses pheromones to lure insects.

Pheromone: A pheromone is a chemical substance that is released by one organism to influence the biology or behavior of another.

Phloem: The vascular tissue in plants, which conducts sugars and other metabolic products downwards from the leaves.

Photosynthesis: It is the process by which plants use the energy from sunlight to produce glucose from carbon dioxide and water.

Phytophagous: An insect or other invertebrate feeding on plants.

Piercing and sucking mouthparts: The defining feature of the order Hemiptera is the possession of mouthparts where the mandibles and maxillae are modified into a proboscis, sheathed within a modified labium, which is capable of piercing tissues and sucking out the liquids. For example, true bugs, such as shield bugs, feed on the fluids of plants.

Pink bud: Stage in the development of the flower of an apple or other fruit tree in which the buds show pink colour and are beginning to open.

Plump: Having a full rounded shape.

Polyvoltine: Those races, which have more than four yearly generations.

Pre-bloom: The time the buds show first colour and the full opening of blossoms.

Predator: A predator is an animal that eats other animals.

Pre-dispose: Make someone liable or inclined to a specified attitude, action, or condition.

Pre-oviposition: The period of time between the emergence of an adult female insect and the start of its egg laying.

Pre-pupae: A larval insect in the stage just preceding pupation.

Progeny: A descendant or the descendants of a person, animal, or plant; offspring.

Propodosoma: Portion of the podosoma that bears the first and second pairs of legs of a tick or mite.

Protogynes: A female that resembles the male of the same species; a normal female.

Protonymph: The newly hatched form of various mites.

Pruning: Reduce the extent of (something) by removing superfluous or unwanted parts.

Pupa: The third stage in the life cycle of insects undergoing complete metamorphosis.

Pupal cells: Underground cells, loose in the soil where a pupa remains.

Puparia: The hardened last larval skin, which encloses the pupa in some insects, especially higher Diptera.

Quiescent: In a state or period of inactivity or dormancy.

Raisin: A partially dried grape.

Receptacle: Thickened part of a stem from which the flower organs grow.

Refuge: A place that provides shelter or protection.

Rind: The tough outer skin of certain fruit, especially citrus fruit.

Russet: Reddish brown in colour.

Saliva: Watery liquid secreted into the mouth by glands, providing lubrication for chewing and swallowing, and aiding digestion.

Sap: The fluid that circulates in the vascular system of a plant, consisting chiefly of water with dissolved sugars and mineral salts.

Sapwood: The soft outer layers of recently formed wood between the heartwood and the bark, containing the functioning vascular tissue.

SC: Soluble concentrate.

Scab: A dry, rough protective crust that forms over a cut or wound during healing.

Scale insect: A small bug with a protective shield-like scale. It spends most of its life attached by its mouth to a single plant, sometimes occurring in such large numbers that it becomes a serious pest.

Scarab: A family of beetles that have distinctively shaped antennae and are often metallic in coloration.

Scarring: A mark left on the skin after a surface injury or wound has healed.

Secateur: A garden tool that has two short sharp blades and is used for cutting plant stems.

Sedentary: Tending to spend much time somewhat inactive.

Senescence: Loss of a cell's power of division and growth. It is the growth phase in a plant or plant part (as a leaf) from full maturity to death.

Sepals: Each of the parts of the calyx of a flower, enclosing the petals and typically green and leaf-like.

Serrate: Notched on the edge like a saw.

Setae: A stiff hair-like or bristle-like structure, especially in an invertebrate.

Sever: Divide by cutting or slicing, especially suddenly and forcibly.

Shellac: Shellac is a resin secreted by the female lac bug, on trees

Shrivel: Wrinkle and contract or cause to wrinkle and contract, especially due to loss of moisture.

Shrub: A woody plant, which is smaller than a tree and has several main stems arising at or near the ground.

Silver tip: It is the first stage of growth in fruit trees after dormancy.

Sinuous: Having many curves and turns.

Skeletonization: Skeletonization refers to the final stage of death, during which the last vestiges of the soft tissues have decayed or dried to the point that the skeleton is exposed.

SL: Soluble (liquid) concentrate.

Slant: Slope or lean in a particular direction.

Slender: Having little width in proportion to height or length.

Slit: A long, narrow cut or opening.

Slug: A tough-skinned terrestrial mollusc, which typically lacks a shell and secretes a film of mucus for protection.

Snout: The projecting nose and mouth of an animal/insect.

Sooty fungus: A blackish fungal growth that develops on plant surfaces that have become covered with honeydew secreted by aphids or other insects

Speckle: Mark with a large number of small spots or patches of colour.

Species: Species is one of the seven taxonomic ranks used to classify living organisms. A species is a group of organisms that can breed and produce fertile offspring.

Spermatophores: A spermatophore or sperm ampulla is a capsule or mass containing spermatozoa created by males of various animal species, especially salamanders and arthropods, and transferred in entirety to the female's ovipore during reproduction.

Spherical: shaped like a sphere or ball.

Spiracle: Openings on the thorax and abdomen of insects that allow the insect to breathe.

Spur: A spur is a spike, usually part of a flower. In certain plants, part of a sepal or petal develops into an elongated hollow spike extending behind the flower, containing nectar, which is sucked by insects.

Stalk: The main stem of a herbaceous plant. It is the slender attachment or support of a leaf, flower, or fruit.

Steely: Resembling steel in colour, brightness, or strength.

Stem mother: A sexually produced female that produces parthenogenetically a colony of offspring; especially: an aphid that develops from an overwintering egg and gives rise to the summer generation

Stipple: Mark a surface with numerous small dots or specks.

Stout: Strong and thick.

Stunting: Prevent from growing or developing properly.

Stylet: A small style, especially a piercing mouthpart of an insect.

Styrofoam: A kind of expanded polystyrene used especially for making food containers.

Sub cortex: The portion of the stem between the epidermis and the vascular tissue or bark.

Succulent: Xerophyte having thick fleshy leaves or stems adapted to storing water.

Suctorial: Adapted for sucking.

Supple: Bending and moving easily and gracefully. Making more flexible.

Swab: Clean with a cloth.

Swarm: A large or dense group of flying insects.

Synchronous: Existing or occurring at the same time.

Tan: Tan is a pale tone of brown. The name is derived from tannum (oak bark) used in the tanning of leather.

Tawny: Orange brown or yellowish brown colour.

Tentiform: Resembling or building a nest that resembles a tent in form.

Thorax: One of the three main body parts of an insect.

Thorn: Thorn is a sharp excrescence on a plant, especially a sharp-pointed aborted branch.

Thysanoptera: Fringe- winged; an ordinal term applied to species with four narrow, similar wings, lengthily fringed; mouth parts fitted for puncturing and scraping; metamorphosis incomplete; the thrips.

Tinge: A trace of a colour.

Toxemia: Blood poisoning resulting from the presence of bacterial toxins in the blood.

Translucent: Permitting the passage of light but objects on the other side can't be seen clearly.

Tricho card: Tricho cards are the cards prepared in the laboratory having Corcyra eggs parasitized by egg parasitoid, Trichogramma.

Trichomes: A small hair or other outgrowth from the epidermis of a plant, typically unicellular and glandular.

Tubercle: Tubercle is a general term for a round nodule, small eminence, or warty outgrowth found on external or internal organs of a plant or an animal.

Tufts: Bunch or cluster of small, usually soft and flexible parts, as feather or hairs, attached or fixed closely together at the base and loose at the upper ends.

Turf: Grass and the surface layer of earth held together by its roots.

Twig: A slender woody shoot growing from a branch or stem of a tree or shrub.

Underneath: The part or side of something facing towards the ground; the underside.

Univoltine: Producing one brood in a season and especially a single brood of eggs capable of hibernating.

Vector: A vector is an organism that acts as an intermediary host for a parasite and transfers the parasite to the next host.

Ventral: Relating to the underside of an animal or plant.

Vigour: It is the physical strength and good health.

Virus: A virus is a small infectious agent that replicates only inside the living cells of other organisms. Viruses can infect all types of life forms, from animals and plants to microorganisms, including bacteria and archaea.

Viviparous: Bringing forth live young, which have developed inside the body of the parent.

Voracious: Wanting or devouring great quantities of food.

Water sprout: A vigorous upright shoot from an adventitious or latent bud on the trunk or main branch of a tree.

WDG: Water dispersible granules.

Weevil: Beetles in the superfamily curculionidae. The group includes several pests of economic importance.

WG: Water dispersible granules.

White cap stage: As a nymph of San Jose scale feeds, it secretes a white waxy scale covering called white cap.

Wilting: Wilting is the loss of rigidity of non-woody part of plants. This occurs when the turgor pressure in non-lignified plant cells falls towards zero, as a result of diminished water in the cells. The rate of loss of water from the plant is greater than the absorption of water in the plant.

Wing pads: These are the undeveloped wings of the active pupa of an insect.

Wither: When a plant becomes dry and shriveled.

Woodlots: Small area of trees that can be used as fuel or to provide wood for building things.

Wriggle: To twist from side to side with small quick movements like a worm.

Xylem: Xylem is one of the two types of transport tissue in vascular plants, phloem being the other. The basic function of xylem is to transport water from roots to shoots and leaves, but it also transports some nutrients.

Yellow sticky trap: Yellow Sticky Traps are the traps that capture insects that are attracted to the yellow colour.

Index

Q

R

S

www.ingramcontent.com/pod-product-compliance
Lightning Source LLC
Chambersburg PA
CBHW050518190326
41458CB00005B/1576
* 9 7 8 9 3 8 7 0 5 7 6 7 8 *